Essentials of
CHEMISTRY

Essentials of CHEMISTRY

Michael L. Maas
Long Beach City College

WM. C. BROWN COMPANY PUBLISHERS
Dubuque, Iowa

To C.H. and E.A. for all they have done.

Contents

Preface

This textbook has been designed for a one-semester chemistry course for the nonscience major. It is hoped, however, that it can also be used in a prerequisite course for future science majors who have had no previous experience in the study of chemistry.

After working with students in these two types of courses, I have concluded that major emphasis should be placed on the areas of nomenclature and stoichiometry, and that a basic study of atomic structure and bonding should be presented. It is my belief that theory, as an entity in itself, is not important, and that this textbook should provide the fundamental knowledge from which additional sophisticated ideas may be developed.

All mathematical concepts presented here require nothing more than a basic understanding of algebra, and it is the intent that students solve problems by proportions whenever possible.

Departures from the usual textbook in this field are the minimum amount of descriptive chemistry and the elimination of the traditional breakdown of the periodic chart into various groups and families for detailed study. However, an overview of periodic properties as related to structure and bonding is presented.

Although the treatment given to organic chemistry and biochemistry in this book is not complicated and involved, it does give a feeling for the interrelationship of these two areas.

It is with gratitude that I acknowledge the contributions of my colleagues and associates toward the preparation of the manuscript. Special recognition goes to Philip Bruce, Steven Parkin, Dr. Howard

Pinckard, and George Slemmer, my colleagues at Long Beach City College, for their helpful comments and suggestions. I am deeply indebted to Charles Cunningham for reviewing the entire manuscript with keen knowledge and understanding, and with the utmost patience. Finally, I extend my sincere appreciation to my wife Kris for her patience and understanding as well as for her outstanding work in typing both the preliminary editions and final manuscript.

M.L.M.

CHAPTER 1

Foundations of Science and Mathematics

1.1 Introduction

Whenever a course of study is begun, it is usually necessary to define the various areas either by a historical definition or by a well-used phrase. However, neither of these methods is adequate for the beginning of a course in chemistry.

Chemistry is prehistoric. And yet, chemistry is today; and to an even greater extent, it is the future. Chemistry is life. Chemistry is with us in all our activities.

Chemistry is but one small area of science. Through this course, a starting point in scientific study, hopefully, will be established. You, the reader, are facing a challenge. Meet this challenge and you will have discovered a stimulating area of your total education. If you remember the fundamental concepts and are logical in your thinking, success will be yours.

1.2 Basic Properties of Matter and Energy

Terminology is a problem to almost everyone who is starting out in a new field of study. Chemistry is no exception. Many terms must be defined—and learned—so that they can be used correctly in future work.

Science deals with the study of nature. Nature is composed of matter and energy.

Matter is anything that has mass and that occupies space. By this

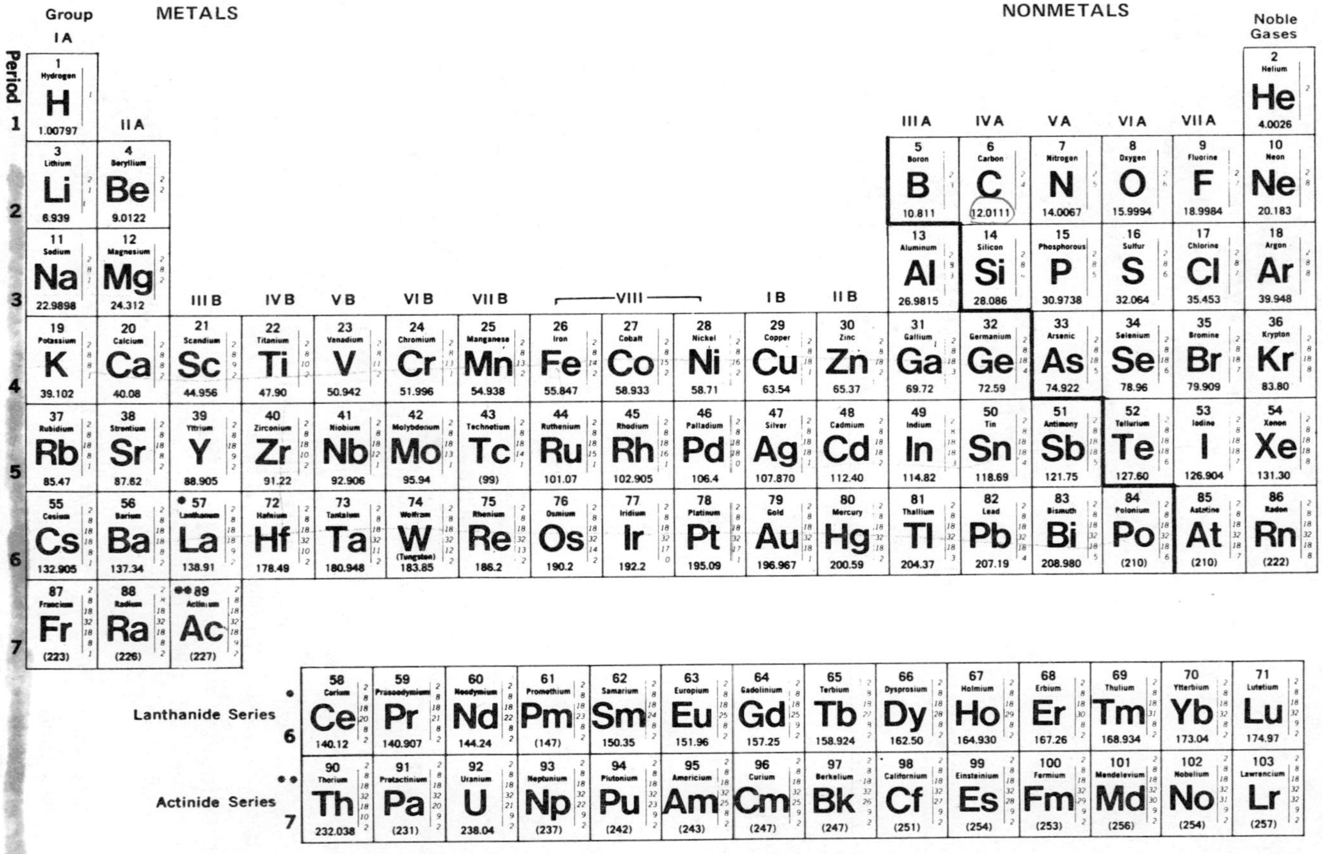

METALS

NONMETALS

Group

Noble Gases

Period

Lanthanide Series

Actinide Series

Figure 1.1. Periodic Chart Illustrating Separation of Metals and Nonmetals.

2

definition, everything other than a vacuum in our world is composed of matter.

Mass is the quantity of matter a body possesses, as evidenced by inertia.

Matter can be divided into very small subgroups, most of which are so small that no person has even seen them. It is these subgroups of matter with which a chemist deals in his work.

Most matter is composed of elements. An element is a pure substance, that is, a substance with a definite and fixed composition. A listing, or arrangement, of these elements according to similar properties is called a *periodic chart.*

A *chemical symbol* is the representation assigned to each of the elements. Each element has its own symbol.

An *atom* is the smallest unit of an element that may participate in a chemical reaction.

A *molecule* is the smallest physical unit of a pure substance. In discussions of chemical reactions, the term *compound* is often used as a synonym for the term molecule.

It has been found to be convenient to divide the periodic chart of the elements into various sections for study. One method is to list elements either as metals or as nonmetals.

A *metal* is an element which illustrates properties such as conductivity, ductility, malleability, and luster. A *nonmetal* is an element which does not exhibit those properties. As is the case with many classifications, this is not a simple, definite, black and white separation. In fact, there is a gray area which lists those elements that have the properties of both metals and nonmetals. An example would be silicon.

To aid this separation, most periodic charts have a line drawn between the metals and the nonmetals. Approximately three-fourths of the elements are metals, and one-fourth are nonmetals. There are many other methods of grouping the elements, some of which will be discussed later.

Energy is the capacity for performing work. All actions in which matter participates involve energy. There are many different forms of energy. Two of the most common forms are called kinetic energy and potential energy.

Kinetic energy is the energy of motion. Examples would be the flight of a bird or the rolling of a marble. Kinetic energy involves both mass and velocity. The derived relationship for kinetic energy is:

$$KE = 1/2 \ mv^2 \qquad \text{where:} \qquad KE = \text{Kinetic Energy}$$
$$m = \text{mass}$$
$$v = \text{velocity}$$

Potential energy is energy of position. A person sitting on a ledge possesses potential energy which is converted to kinetic energy when the person jumps to the ground. This conversion is a slow, continual process as a body moves from a position of rest to one of movement.

1.3 Physical and Chemical Properties of Matter

In science, substances are identified by their unique properties. It is convenient to divide these properties or characteristics into two groups.

Physical properties are those characteristics which may be determined without changing the composition of the substance. Examples of physical properties would be color, taste, odor, melting point, boiling point, density, and hardness.

Chemical properties are those characteristics which may be determined only through a change in the composition or structure of the substance. Examples of chemical properties would be the burning of gasoline, the rusting of metals, and the decomposition of wood.

The study of changes in physical and chemical properties gives rise to the basic concept of chemical reactions. A *chemical reaction* is the changing of one or more substances into one or more new substances which have different chemical and physical properties than did the original species. Thus, by understanding properties, it will be much easier to determine whether a chemical reaction, or merely a change of state has occurred. The changes of state for water are illustrated by Figure 1.2. Figure 1.3 shows the chemical reaction of burning sugar in air.

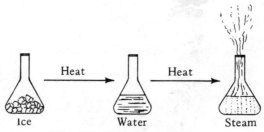

Figure 1.2. Three States of Water.

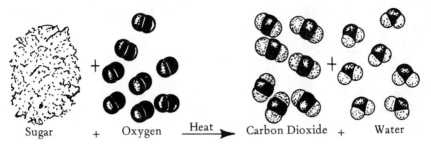

Sugar   +   Oxygen $\xrightarrow{\text{Heat}}$ Carbon Dioxide   +   Water

Figure 1.3. The Burning of Sugar.

1.4 Mathematical Concepts

It is often necessary to work with very large or with very small numbers. For example, the diameter of a molecule might be 0.000000036 cm, while the number of molecules in a milliliter of gas could be equal to 20,000,000,000,000,000,000. From these examples alone it can be noted that it would be very time consuming as well as impractical to write such numbers in this form every time a calculation is necessary. It is possible to write very small and very large numbers in a more convenient form. This form is called *exponential form.*

The use of exponents is very convenient in our decimal system because it is based on the number 10. We know, of course, that the number 10 is ten times larger than the number 1; the number 100 is ten times larger than the number 10; the number 1,000 is ten times larger than the number 100 and so on. To illustrate, study the powers of ten given in Table 1.1.

TABLE 1.1
EXPONENTIAL NOTATION FOR LARGE NUMBERS

$$10 = 10^1$$
$$100 = 10 \times 10 = 10^2$$
$$1000 = 10 \times 10 \times 10 = 10^3$$
$$10000 = 10 \times 10 \times 10 \times 10 = 10^4$$
$$100000 = 10 \times 10 \times 10 \times 10 \times 10 = 10^5$$
$$1000000 = 10 \times 10 \times 10 \times 10 \times 10 \times 10 = 10^6$$

When 10 is used as a factor twice, the exponent is 2. When 10 is used as a factor five times, the exponent is 5. Notice the relationship between the number of zeros in a number and the exponent. How would you express 100,000,000 as a power of ten?

This same procedure can be followed in working with small numbers. By following the pattern established in Table 1.1, it is evident that $1 = 10°$. This is a basic law of mathematics. That is, any number expressed to the zeroth power is equal to one.

When numbers less than one are considered, negative exponents come into existence. The decimal fraction 0.1 is 10 times smaller than the number 1, the fraction 0.01 is ten times smaller than the fraction 0.1 and so on. To illustrate, study the relationships shown in Table 1.2.

TABLE 1.2
EXPONENTIAL NOTATION FOR VERY SMALL NUMBERS

$$1 = 10^0$$
$$.1 = 1/10 = 10^{-1}$$
$$.01 = 1/10 \times 1/10 = 10^{-2}$$
$$.001 = 1/10 \times 1/10 \times 1/10 = 10^{-3}$$
$$.0001 = 1/10 \times 1/10 \times 1/10 \times 1/10 = 10^{-4}$$
$$.00001 = 1/10 \times 1/10 \times 1/10 \times 1/10 \times 1/10 = 10^{-5}$$

Thus far, a convenient method has been established for the writing of both very large and very small numbers. However, it is obvious that on most occasions the number involved will not be a perfect multiple of ten. Therefore, a method for working with any large or small number is necessary.

For example, consider the number 6,000,000. By rewriting this number as $6 \times 1,000,000$ it could very easily be fitted into the above system as 6×10^6. A number expressed in this way is said to be written in *exponential notation*. A number is expressed in exponential notation if it is written as a product of a number between one and ten and the proper power of ten. If a number is a perfect power of ten, the first factor is one and often is not indicated. Study Table 1.3 and notice how both very large and very small numbers may be expressed in exponential notation.

The mathematical operations of multiplication and division become much easier when working with numbers expressed in exponential notation. Consider the example:

$$5,000 \times 7,300 = 36,500,000$$
$$5 \times 10^3 \times 7.3 \times 10^3 = 36.5 \times 10^6 = 3.65 \times 10^7$$

The algebraic rule for the multiplication of numbers in exponential form is to multiply the coefficients and algebraically to add the exponents.

TABLE 1.3
EXPONENTIAL NOTATION

Number	Exponential Notation
18	1.8×10^1
62.3	6.23×10^1
193.7	1.937×10^2
4,000	4×10^3
230,000	2.3×10^5
63,000,000	6.3×10^7
0.4	4×10^{-1}
0.0032	3.2×10^{-3}
0.00007	7×10^{-5}
0.00000051	5.51×10^{-7}

Table 1.4 illustrates additional examples of the multiplication of exponential numbers. The parentheses are used to indicate a multiplication operation.

TABLE 1.4
MULTIPLICATION IN EXPONENTIAL NOTATION

$$(2 \times 10^3)(3 \times 10^5) = 6 \times 10^8$$
$$(4.5 \times 10^5)(8 \times 10^{10}) = 3.6 \times 10^{16}$$
$$(5 \times 10^{23})(10^{15}) = 5 \times 10^{38}$$
$$(2.5 \times 10^{10})(10^{-15}) = 2.5 \times 10^{-5}$$
$$(6 \times 10^6)(5 \times 10^{-3}) = 3 \times 10^4$$
$$(8 \times 10^{-4})(1 \times 10^{-3}) = 8 \times 10^{-7}$$

Since division is the opposite operation of multiplication, just the opposite operation is required mathematically. The algebraic rule for dividing numbers in exponential form is to divide the coefficients and algebraically to subtract the exponents. Study Table 1.5 for examples of division of exponential numbers.

Anytime a measurement is made, be it in science, mathematics or just plain everyday living, there is an amount of uncertainty associated with it. Suppose a student were asked to measure the width of a classroom. One student might use a ruler calibrated to the nearest inch and obtain a width of 41 feet 6 inches. In turn, another student might use an uncalibrated yard stick and obtain a value of 14 yards. Still another student might use a ruler with units to the nearest 0.1 inch and obtain a value of 41 feet 6.3 inches. From this illustration,

TABLE 1.5
DIVISION IN EXPONENTIAL NOTATION

$$\frac{10^5}{10^2} = 10^{5-2} \qquad = 10^3$$

$$\frac{6 \times 10^4}{2 \times 10^3} = 3 \times 10^{4-3} \qquad = 3 \times 10^1$$

$$\frac{2.1 \times 10^8}{3 \times 10^5} = .7 \times 10^{8-5} \qquad = 7 \times 10^2$$

$$\frac{9 \times 10^{10}}{3 \times 10^{15}} = 3 \times 10^{10-15} \qquad = 3 \times 10^{-5}$$

$$\frac{8 \times 10^{15}}{4 \times 10^{-10}} = 2 \times 10^{15-(-10)} = 2 \times 10^{25}$$

it can be noted that measurement cannot be considered exact, but rather approximate.

However, it can be stated that the smaller the unit of measure, the greater is the *precision.* In the above illustration, the measurement with the ruler in units of 0.1 inch would be the most precise or in other words, have the greatest precision.

Precision has another meaning also and that is in laboratory work where the precision is a measure of the reproducibility of an answer.

Remember though, every measurement has a degree of uncertainty. The *greatest possible error* of a measurement is equivalent to one-half the unit of measure being used. Table 1.6 shows the relationship between precision and uncertainty or greatest possible error.

TABLE 1.6
PRECISION AND UNCERTAINTY OF MEASUREMENT

Measurement	Precision	Greatest Possible Error
41 ft 6 in	1 in	½ in
14 yds	1 yd	½ yd
41 ft 6.3 in	0.1 in	0.05 in

Since any measurement is only an approximation, it is important to know just how exact or how accurate the measurement really is.

Two measurements may be made with the same precision; however, the error of measurement may be more serious in one case than the other. We can obtain a measure of the importance of the greatest possible error by comparing it with the actual measurement.

The measurements 25 feet and 5 feet are made with the same precision. If the greatest possible error in each measurement is divided by the actual measurement, the result would be:

$$\frac{0.5}{25} = 0.02 = 2\%$$

$$\frac{0.5}{5} = 0.10 = 10\%$$

These quotients, 2% and 10% respectively, are called *relative errors*. Relative error can be found by dividing the greatest possible error by the actual measurement.

The smaller the relative error, the greater is the *accuracy* of the measurement. Relatively, the measurement of 25 feet is more accurate than the measurement of 5 feet because the relative error or per cent of error is less in 25 feet than in 5 feet.

It is apparent that it would be rather time consuming to have to determine the accuracy of a measurement by this method in all experimental work. Fortunately, the accuracy of a measurement can be determined without calculation (See Table 1.7). The method used to determine accuracy of a measurement is by counting the number of *significant digits* in a numeral. Very simply, the more significant digits in a numeral of measurement, the more accurate is the measurement. Remember, the accuracy of a measurement is independent of the units of measure used.

There are rules for determining the number of significant digits in a numeral:

1. Every non-zero digit is significant.
2. Every zero between non-zero digits is significant.
3. Any zero to the right of the decimal point *and also* on the extreme right of the numeral is significant.

Table 1.7 summarizes the concepts of precision, accuracy, and significant digits. Note the relationship between relative error and the number of significant digits.

It is essential that the above concepts be used in all calculations. No matter what the mathematical operation, remember that the derived quantity can have only as many significant digits as the least accurate quantity from which it was calculated. For example, (236 ft) (1,000 ft) = 236,000 ft² = 200,000 ft² . Your answer should have

TABLE 1.7
PRECISION, ACCURACY AND SIGNIFICANT DIGITS

Measurement	Precision	Greatest Poss. Error	Rel. Error	Number of Sig. Digits
125 lb	1 lb	.5 lb	0.4%	3
2500 yd	100 yd	50 yd	2%	2
20.15 mi	.01 mi	.005 mi	0.02%	4
0.0005 ft	.0001 ft	.00005 ft	10%	1
0.0105 in	.0001 in	.00005 in	0.5%	3

but one sigificant digit, as the factor 1,000 contains but one significant digit.

From the above example, it is apparent that some measurements have an uncertainty as to the number of significant digits. The measurement 1,000 ft could have one, two, three, or four significant digits depending upon the individual who measured. Therefore, to eliminate any doubt, write the measurement in exponential notation which will include only the significant figures.

1.5 The Metric System

Since the metric system of measure is based on ten and powers of ten, exponential notation is especially useful in dealing with metric units. Due to the relationship between the metric system and the base ten number system, the metric system of measure has been used by scientists for many years.

The basic unit of linear measure is the *meter* which was originally defined as one ten-millionth of the distance from the North Pole to the equator along a meridian. In 1960, the meter was redefined in terms of the orange-red wavelength of radiating krypton–86 gas. In terms of our English system of measures, a meter is a little over one yard.

In Table 1.8 the units of linear measure of the metric system are summarized. Notice the relationship between the various units as you study the figure.

Notice that each metric unit in the first column uses the word *meter* with a prefix. The meaning of each prefix is given by the corresponding numeral in the third column. In other words, *milli* means 1/1000; *centi* means 1/100 and so on. These prefixes will have the same meaning no matter what type of unit of measure is being used.

TABLE 1.8
LINEAR UNITS OF MEASUREMENT IN THE METRIC SYSTEM

Unit	Abbreviation	Eq. in Meters	Ex. Notation
1 millimeter	1 mm	1/1000 m	10^{-3} m
1 centimeter	1 cm	1/100 m	10^{-2} m
1 decimeter	1 dm	1/10 m	10^{-1} m
1 meter	1 m	1 m	1 m
1 decameter	1 dkm	10 m	10^{1} m
1 hectometer	1 hm	100 m	10^{2} m
1 kilometer	1 km	1000 m	10^{3} m

It should be pointed out that some of these metric units of length are used much more often than others and for this reason will be emphasized more than the others. The basic units of length used in this course will be the millimeter, centimeter, meter, and kilometer.

There is one additional unit of length used in chemistry to measure very small distances and it also is based on the number 10. An *Ångström* (Å) is equivalent to 10^{-8} cm or 10^{-10} meters.

Although the metric system of measures is the official system, both the metric system and the English system of measures are used in the United States. Therefore, it is necessary to be able to convert measurements from one system to the other as well as to perform conversions within either system. Table 1.9 lists some of the common conversion units of length between the metric system and the English system of measures. It should be noted, these conversions are approximate values rounded to the significance shown.

TABLE 1.9
CONVERSION UNITS OF LENGTH

English Unit	Metric Unit
1 in	2.54 cm
39.37 in	1 m
0.62 mi	1 km

Study the following examples to become familiar with the mathematics involved in changing from one unit of measure to another unit of measure in the same system.

Example 1.1: Convert 5450 meters to centimeters.

$$\frac{x \text{ cm}}{100 \text{ cm}} = \frac{5450 \text{ meter}}{1.00 \text{ meter}}$$

$$x = \frac{(100 \text{ cm})(5450 \text{ meters})}{1.00 \text{ meter}}$$

$$x = 545000 \text{ cm} = 5.45 \times 10^5 \text{ cm}$$

Example 1.2: Convert 5×10^7 mm to meters.

$$\frac{x \text{ meter}}{1 \text{ meter}} = \frac{5 \times 10^7 \text{ mm}}{10^3 \text{ mm}}$$

$$x = \frac{(1 \text{ meter})(5 \times 10^7 \text{ mm})}{10^3 \text{ mm}}$$

$$x = 5 \times 10^4 \text{ meter}$$

Example 1.3: Convert 25 km to cm.

$$\frac{x \text{ cm}}{10^5 \text{ cm}} = \frac{25 \text{ km}}{1.0 \text{ km}}$$

$$x = \frac{(25 \text{ km})(10^5 \text{ cm})}{1.0 \text{ km}}$$

$$x = 25 \times 10^5 \text{ cm} = 2.5 \times 10^6 \text{ cm}$$

From the above examples, it is evident how useful exponential notation and the ability to solve proportions can be in carrying out unit conversions. All conversions can be done in this same manner of setting up a proportion and cross-multiplying to obtain the solution.

The next set of examples show how to convert from one system of measurement to a different system of measurement.

Example 1.4: Convert 8.00 inches to centimeters.

$$\frac{x \text{ cm}}{2.54 \text{ cm}} = \frac{8.00 \text{ inches}}{1.00 \text{ inches}}$$

$$x = \frac{(2.54 \text{ cm})(8.00 \text{ inches})}{1.00 \text{ inches}}$$

$$x = 20.32 \text{ cm} = 20.3 \text{ cm} \text{ (3 significant digits)}$$

Example 1.5: Convert 3×10^4 meters to inches.

$$\frac{x \text{ inches}}{39.37 \text{ inches}} = \frac{3 \times 10^4 \text{ meters}}{1 \text{ meter}}$$

$$x = \frac{(39.37 \text{ inches})(3 \times 10^4 \text{ meters})}{1 \text{ meter}}$$

$$x = 118.11 \times 10^4 \text{ inches} = 1 \times 10^6 \text{ inches}$$

(1 significant digit)

Example 1.6: Convert 20 feet to centimeters.

Before setting up the proportion, change 20 feet to an equivalent number of inches as the conversion factor between inches and centimeters is known from Table 1.9. (20 ft) (12 in/ft) = 240 inches.

$$\frac{x \text{ cm}}{2.54 \text{ cm}} = \frac{240 \text{ in}}{1 \text{ in}}$$

$$x = \frac{(2.54 \text{ cm}) (240 \text{ in})}{1 \text{ in}}$$

$$x = 609.60 \text{ cm} = 610 \text{ cm} = 600 \text{ cm}$$
$$\text{(1 significant digit)}$$

Just as in the English system of measures, the unit for measuring volume in the metric system is the volume of a cube. If a cube has an edge 1 meter in length, the volume would be found by multiplying the length times the width times the height or in other words, (1 meter) (1 meter) (1 meter) = 1 *cubic meter.*

An even smaller cube could have an edge with a length of 1 centimeter. Thus, the volume would be 1 cubic centimeter (cc). Such a cube would be about the size of an ordinary sugar cube, about 0.4 inch on each edge. Now let us consider a cubical container with an edge of 1 decimeter or 10 centimeters as shown below. The volume of a cube with an edge of 10 cm would be (10 cm) (10 cm) (10 cm) or 1000 cc.

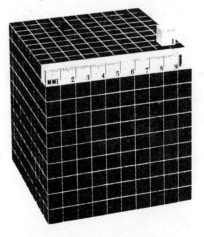

Figure 1.4. A Cube With an Edge of 10 Centimeters.

Often, the term *capacity* is used with respect to the discussion of volume. To be specific, capacity simply means the total volume of a container. For example, when a pail holds one gallon of water (capacity) we are indirectly saying that its volume is 231 cu in.

The cube in Figure 1.4 has a volume of one cubic decimeter or 1000 cc. The capacity of such a container is exactly one *liter* which is the usual unit of capacity in the metric system.

As a point of reference, the liter and the quart are approximately the same size. To be specific, one quart is equal to 0.946 liters or 946 milliliters.

Two other measures of capacity are the *milliliter* and the *kiloliter* which are related to the liter as indicated in the statements below.

$$1 \text{ milliliter (ml)} = 0.001 \text{ liter} = 1 \times 10^{-3} \text{ liters}$$

$$1 \text{ kiloliter (kl)} = 1000 \text{ liters} = 1 \times 10^{3} \text{ liters}$$

Since 1 liter corresponds to 1000 cc; then 0.001 liter corresponds to 1 cc. Furthermore, 1 milliliter of water equals 1 cc of water.

Example 1.7: Convert 3,000 cc to milliliters and liters.

$$1 \text{ cc} = 1 \text{ ml} \quad \text{Therefore, } 3{,}000 \text{ cc} = 3{,}000 \text{ ml}$$

$$\frac{x \text{ liters}}{1 \text{ liter}} = \frac{3{,}000 \text{ ml}}{1{,}000 \text{ ml}}$$

$$x = \frac{(3{,}000 \text{ ml})(1 \text{ liter})}{1{,}000 \text{ ml}} = 3 \text{ liters}$$

The last area of study in measurement is that of mass and weight. Mass and weight do not mean exactly the same thing.

The *mass* of an object does not change with, nor does it depend on, the position of the object in space where it is measured.

The *weight* of an object does change with, and does depend on, the position in space where it is measured. Weight deals with the acceleration of gravity while mass does not.

The metric unit of 1 *gram* (1 gm) is defined as the mass of 1 cubic centimeter of water at 4°C. This definition is very convenient because it states a definite relationship between mass and volume or capacity. Table 1.10 summarizes the concepts of volume, capacity and mass as they have now been defined with respect to water.

At times, it is necessary to make conversions from the metric system of mass to the English system. The conversion factors are:

$$1.00 \text{ kilogram} = 2.20 \text{ lbs}$$

$$454 \text{ grams} = 1.00 \text{ lb}$$

TABLE 1.10
THE RELATIONSHIP BETWEEN VOLUME,
CAPACITY AND MASS OF WATER

Volume	Mass	Capacity
1 cc	1 gm	1 ml
1 dm^3	1 kg	1 liter
1 m^3	1 metric ton	1 kiloliter

Example 1.8: Convert 2724 grams to pounds.

$$\frac{x \text{ lbs}}{1.00 \text{ lbs}} = \frac{2724 \text{ grams}}{454 \text{ grams}}$$

$$x = \frac{(1.00 \text{ lbs}) (2724 \text{ grams})}{454 \text{ grams}} = 6.00 \text{ lbs}$$

1.6 Temperature Scales

Temperature is a method for determining the spontaneous direction of heat flow. In science, there are three temperature scales that are normally used. They are the *Celsius* or centigrade scale, the *Fahrenheit* scale, and the *Kelvin* scale. It would be convenient if all three scales were the same. Unfortunately, this is not the case. The Celsius and Kelvin degrees are the same size but the Fahrenheit degree is 5/9 as large as the above two. In addition, reference points such as the freezing and boiling points of water are different on all three scales. Therefore, the student of science must be capable of converting from one temperature scale to another. Figure 1.5 illustrates the three temperature scales with respect to the freezing and boiling points of water.

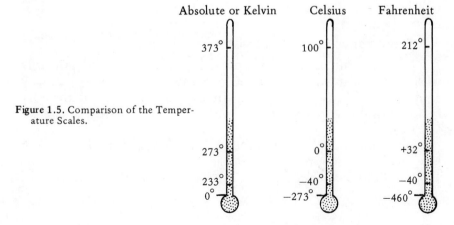

Figure 1.5. Comparison of the Temperature Scales.

From the three scales, the freezing point of water is found to be 0°C, 32°F, and 273°K. In the same manner, the boiling point of water is found to be 100°C, 212°F, and 373°K. By working with these relationships mathematically, formulas for temperature conversion may be obtained:

$$°K = °C + 273$$
$$°C = 5/9(°F - 32)$$
$$°F = (9/5 × °C) + 32$$

Example 1.9: Convert 41°F to °C and °K.

First change to °C.

$$°C = 5/9(°F - 32)$$
$$°C = 5/9(41 - 32) = 5/9(9) = \underline{5° \text{ C}}$$
$$°K = °C + 273 = 5 + 273 = \underline{278° \text{ K}}$$

Example 1.10: Convert −4°C to °K and °F.

$$°K = °C + 273 = (-4) + 273 = 269° \text{ K}$$
$$°F = 9/5 °C + 32$$
$$°F = 9/5 (-4) + 32 = (-7.2) + 32 = 24.8°F$$

Example 1.11: Convert 298°K to °C and °F.

By rearranging the relationship between Kelvin and Celsius:

$$°C = °K - 273$$
$$°C = 298 - 273 = 25° \text{ C}$$
$$°F = 9/5 °C + 32$$
$$°F = 9/5 (25) + 32 = 45 + 32 = 77°F$$

1.7 Fundamental Laws of Chemistry

To this point in the course we have been concerned primarily with terminology and the basic operations necessary to actually study chemistry. Now we are about to consider some of the very basic principles or concepts upon which much of the field of chemistry is based.

In section 1.2 physical and chemical changes were discussed. A law which is closely related to physical and chemical processes is the Law of Conservation of Mass. The *Law of Conservation of Mass* states that in all processes, the total mass of the substances involved remains unchanged. They may change in form, but the total mass is constant in a closed system. Thus, the total mass of the reactants must equal the final mass of the products after the completion of a chemical reaction. This concept is basic to all calculations based on chemical equations. As will be noted in chapter nineteen, the *Law of Conservation of Mass* does not apply to nuclear reactions.

Another law which has been accepted just about as long as there has been a study of chemistry is the Law of Constant Composition. The *Law of Constant Composition* states that the relationship between the masses of the various materials in a pure substance is always the same. If this were not the case, a given substance would have a variety of properties depending on how it had combined. It is obvious that when we discuss a water molecule we must at all times have exactly the same molecule no matter where it has come from.

In close association with the Law of Constant Composition is the Law of Definite Proportions. The *Law of Definite Proportions* states that in any chemical reaction, there is a definite ratio between the mass of each reactant or product and the mass of any other reactant or product.

As an illustration, consider the burning of sulfur in pure oxygen. Figure 1.6 shows that the ratio remains constant even though the actual amounts of reactants and products may vary.

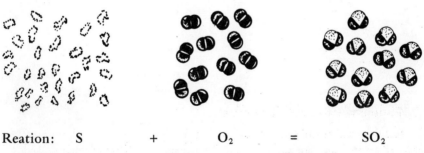

Reation: S   + O_2 = SO_2

Sulfur + oxygen gas = Sulfur dioxide

Example A: 50 gm + 50 gm = 100 gm

Example B: 200 gm + 200 gm = 400 gm

Figure 1.6 Illustration of the Law of Definite Proportions.

The mass ratio between sulfur and oxygen gas is a 1:1 ratio and remains the same in both reactions. The mass ratio between sulfur and sulfur dioxide is 1:2 and also remains the same in both reactions. Whenever sulfur and oxygen gas are combined, the mass ratio for the reaction will be constant.

A final law to consider at this time is the Law of Simple Multiple Proportions. The *Law of Multiple Proportions* states that the same elements may combine in different ratios to form two or more different compounds. This is always the case. That is, if there exists a series of two or more compounds containing the same elements, there is a ratio of small whole numbers between the different weights of one element and a given weight of the other element.

As an example of this law, consider the reaction of hydrogen and oxygen to form water in one case and hydrogen peroxide in the other case. In each case, a weight ratio between oxygen and hydrogen is determined. Study Figure 1.7. Notice that the weight ratio is 1:8 in the formation of water and 1:16 in the formation of hydrogen peroxide. By comparing these two weight ratios, it is apparent that one is exactly twice the other.

$$\underset{\text{2 gm.}}{\text{Hydrogen}} + \underset{\text{16 gm.}}{\text{Oxygen}} = \underset{\text{18 gm.}}{\text{Water}} \qquad \text{Ratio: } \frac{2}{16} = 1{:}8$$

$$\underset{\text{2 gm.}}{\text{Hydrogen}} + \underset{\text{32 gm.}}{\text{Oxygen}} = \underset{\text{34 gm.}}{\text{Hydrogen Peroxide}} \qquad \text{Ratio: } \frac{2}{32} = 1{:}16$$

Figure 1.7. Illustration of the Law of Multiple Proportions.

Glossary

Ångström—A unit of length equal to 1×10^{-8} centimeters.

Atom—The smallest unit of an element that may participate in a chemical reaction.

Chemical Properties—Characteristics of matter which may be determined only through a change in the composition of the substance.

Chemical Reaction—The changing of one or more substances into new substances which have different chemical and physical properties than the original species.

Chemical Symbol—The representation assigned each of the elements. Each element has its own symbol.

Compound—A chemical combination of two or more elements.

Energy—The capacity for performing work.

Element—A substance not separable by ordinary chemical means into substances different from itself.

Kinetic Energy—The energy of motion. Kinetic energy is equal to one-half the mass times the velocity squared of any moving object.

Law of Conservation of Mass—In normal physical and chemical processes the total mass of the substances involved remains unchanged.

Law of Constant Composition—The relationship between the mass of materials in a pure substance is always the same.

Law of Definite Proportions—In a chemical reaction there is a definite ratio between the mass of each reactant or product and the mass of any other reactant or product.

Law of Multiple Proportions—The same elements may combine in different ratios to form two or more different compounds.

Mass—The quantity of matter a body possesses as evidenced by inertia.

Matter—Anything that has mass and occupies space.

Metal—Any substance which illustrates such metallic properties as conductivity, ductility, malleability, and luster.

Molecule—The smallest physical unit of a pure substance.

Nonmetal—Any substance which does not exhibit the metallic properties.

Periodic Chart—A listing or arrangement of the elements according to common chemical and physical properties.

Physical Properties—Characteristics which may be determined without changing the composition of the substance.

Potential Energy—Energy which a body possesses by virtue of its position relative to a frame of reference.

Exponential Notation—A system in which a number is written as a product of a number between one and ten and the proper power of ten.

Weight—The force with which a given body is attracted to a massive body. The acceleration due to gravity on earth is 980 cm/sec^2 or 32 ft/sec^2.

Exercises

1. Explain the difference between the terms *mass* and *weight*. Give an example to illustrate this difference.

2. Classify each of the following properties as chemical or physical: odor, the ability to support combustion, color, solubility.

3. Give three examples of physical changes in matter.

4. Give three examples of chemical changes in matter.

5. Determine the number of significant digits in the following numbers:

   a. 103
   b. 3,000
   c. 1,003.05
   d. 0.003
   e. 10.040

6. Write the following numbers in proper exponential notation.

   a. 7,600,000
   b. 3.06,000,000,000
   c. 1,420,000
   d. 0.000013
   e. 0.0000002405

7. Perform the indicated operations expressing your answer in exponential notation.

   a. $(2.3 \times 10^{-6})(3.1 \times 10^{12}) =$

   b. $(6 \times 10^{23})(2.5 \times 10^{12}) =$

   c. $\dfrac{5.6 \times 10^{10}}{8 \times 10^3} =$

   d. $\dfrac{4.9 \times 10^{-14}}{7 \times 10^{-9}} =$

   e. $\dfrac{(2.31 \times 10^{10})(9.6 \times 10^{-6})}{(2.5 \times 10^{18})} =$

8. Perform the following conversions:

   a. 1.00 cm = _____ in

   b. 3.10 m = _____ ft

   c. 45 km = _____ mi

   d. 32 yd = _____ m

   e. 144 in = _____ cm

9. Complete the following table:

millimeters	centimeters	meters	kilometers
_____	224	_____	_____
_____	_____	_____	1.5
_____	_____	455	_____
_____	1456	_____	_____
9.6×10^8	_____	_____	_____

10. Complete the following table:

milligrams	grams	kilograms
7.25×10^9	_____	_____
_____	496	_____
_____	_____	1.6
3.68	_____	_____

11. Complete the following table:

milliliters	liters	cubic centimeters
_____	2.55	_____
_____	_____	1940
3550	_____	_____
_____	_____	5.2×10^5

12. Complete the following table:

inches	feet	meters	centimeters	millimeters
520	___	___	_____	_____
_____	___	186	_____	_____
_____	14	___	_____	_____
_____	___	___	_____	2.1×10^6
_____	___	___	668	_____

13. Complete the following table:

°C	°K	°F
25	___	___
150	___	___
___	300	___
___	___	68
___	12	___

14. Complete the following table:

measurement	precision	rel. error	significant digits
135 cm	_____	_____	_____
10 in	_____	_____	_____
0.012 mm	_____	_____	_____

15. Rank the following measurements in terms of accuracy. That is, express the most accurate first and so on down to the least accurate. 25 m, 130.0 cm, 5,000 mi, 32.004 in, 5280 ft.

16. A beam of sodium arc light has a wavelength of 6×10^{-5} cm. Express this answer in Ångströms.

17. What are the units for the value of the kinetic energy, $KE = \frac{1}{2}mv^2$, if m is expressed in grams and v in centimeters per second?

18. Density is defined as the mass of an object divided by the volume the substance occupies. That is, $D = m/v$. Determine the density of a metal if 5.7 grams of the metal occupies a volume of 1.9 ml. Express the answer in g/1.

19. What volume of water is occupied by 234 grams of water at 4°C?

20. What volume is occupied by 10.00 g of a substance if the density of the substance is known to be 5.62 g/ml?

CHAPTER 2

Introduction to Formulas and Equations

2.1 Atoms—The Fundamental Particles of Matter

In the previous chapter, an atom was defined to be the simplest particle into which an element may be separated without losing its distinct chemical properties. In further detail, an atom is composed of a fixed number of electrons, protons, and a variable number of neutrons. The electron, proton, and neutron are often referred to as subatomic particles. That is, they are the particles which combine together in a variety of ways to give rise to all the elements.

It is known that the electron has a negative charge, the proton a positive charge, and the neutron is neutral. When combined together, these particles give rise to a neutral atom. All atoms are electrically neutral. Greater study of atomic structure will be discussed in chapter three. It is sufficient at this time to remember that atoms are the simplest particles of matter composed of electrons, protons, and neutrons having a neutral charge.

2.2 Molecules—The Chemical Combination of Atoms

From a great deal of experimentation and theoretical study, it has been found that very few atoms will remain by themselves for a very long period of time. Rather, atoms will combine together to form molecules. The chemical properties of the various atoms dictate how they combine together to form the various molecules. Figure 2.1 illustrates the relationship between atoms and molecules.

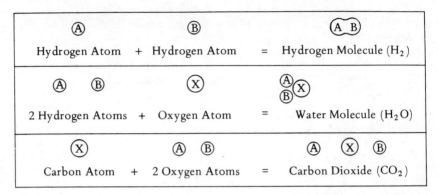

Figure 2.1. Relationship Between Atoms and Molecules.

2.3 Writing Chemical Symbols and Molecular Formulas

Each element has a unique representation which is referred to as a *chemical symbol*. A chemical symbol for each element has been determined when it was discovered and is composed of one or two letters. At times, there may seem to be an inconsistancy in the symbols, but it should be noted that many of the symbols are derived from languages such as Latin, Greek, German, and Russian. Table 2.1 lists symbols for some of the elements. A complete list may be found on the inside front cover of the textbook.

TABLE 2.1
ATOMIC SYMBOLS

Element	Symbol	Element	Symbol
Hydrogen	H	Sodium	Na
Helium	He	Iron	Fe
Carbon	C	Chlorine	Cl
Nitrogen	N	Silver	Ag
Oxygen	O	Copper	Cu

Notice when there is but one letter in the symbol it is a capital letter. When there are two letters in the symbol, the first letter is a capital letter and the second a lower-case letter. Always be consistent and neat when writing chemical symbols so there is no question as to which element you are referring to or writing.

A chemical formula is the representation given to a molecule. The formula is written by combining the chemical symbols for the atoms which compose the molecule. Subscripts are used to indicate the number of the atoms in a molecule with the number one being understood and not written. Table 2.2 illustrates the correct formulas for some common chemical compounds.

TABLE 2.2
CHEMICAL FORMULAS

Compound	Chemical Formula	Compound	Chemical Formula
Water	H_2O	Ammonia	NH_3
Carbon dioxide	CO_2	Nitrogen dioxide	NO_2
Hydrogen gas	H_2	Sulfur trioxide	SO_3
Sodium chloride	$NaCl$	Carbon monoxide	CO
Oxygen gas	O_2	Carbon tetrachloride	CCl_4

Consider the chemical formula for water. In words, the chemical formula states that one molecule of water is composed of two atoms of hydrogen and one atom of oxygen. All chemical formulas can be interpreted in much the same manner.

The question as to which symbol is written first is one of some confusion to beginning students. There is agreement on the part of chemists that, in general, the more electronegative element is written last. That is, write the symbol of the element which is on the left side of the periodic chart and then follow it with the symbol of the element on the right side of the chart. There are exceptions, but this is a general rule to follow.

2.4 Determining Atomic and Molecular Masses (Weights)

The approximate atomic mass or atomic weight of a given element can be determined by referring to the periodic chart and simply reading the average atomic mass number. For example, the average atomic mass of a hydrogen atom is 1.0080 atomic mass units or amu's. By definition, 1 amu is exactly 1/12 the mass of the carbon-12 atom. Due to the number of significant figures in the atomic mass, the mass value is often rounded to two or three significant figures to correspond with respect to precision and accuracy for most calculations. You will notice that the atomic mass of an element listed on the periodic chart is not a whole number, but rather a repeating decimal. The reason for this is that there are various isotopes of each element and the masses are averaged together to obtain the average atomic mass listed on the periodic chart.

The molecular mass or molecular weight of a molecule is determined by summing the atomic masses of the atoms which compose the given molecule. For example, the molecular mass of water, H_2O, is found by adding the mass of two hydrogen atoms and one oxygen atom.

$$2 \text{ H atoms} = \ (2)\,(1.008) = \ 2.016 \text{ amu}$$
$$\underline{1 \text{ O atom} = (1)\,(15.999) = 15.999 \text{ amu}}$$
$$\text{molecular mass } H_2O = 18.015 \text{ amu} = 18.0 \text{ amu's}$$

Table 2.3 summarizes the concepts of atomic mass and molecular mass determinations. Study and be capable of doing such calculations.

TABLE 2.3
ATOMIC AND MOLECULAR MASS

Element	Atomic Mass (amu)	Molecule	Molecular Mass (amu)
H	1.0080	H_2O	18.0
C	12.0112	CO_2	44.0
N	14.0067	NO_2	46.0
O	15.9994	CO	28.0
Na	22.9898	N_2	28.0

2.5 Determining Chemical Composition

After calculating the atomic mass or atoms and the molecular mass of molecules, a logical development is to determine the *percentage composition* of a molecule. Percentage composition for a molecule is determined by expressing each element as a weight percentage of the total molecular mass.

Example 2.1: Determine the percentage composition for water, H_2O.

Step 1: Determine the molecular mass of H_2O.

$$\text{hydrogen} = 2H = (2)\,(1.0) = \ 2.0 \text{ amu}$$
$$\underline{\text{oxygen} = \ O = \ (1)\,(16) = 16.0 \text{ amu}}$$
$$\text{molecular mass } H_2O \ \ \ = 18.0 \text{ amu}$$

Step 2: Determine the weight percentages.

a. $\% H = \dfrac{\text{mass H}}{\text{molecular mass}}\,(100) = \dfrac{2.0}{18.0} = (100) = 11.1\%$

b. $\% O = \dfrac{\text{mass O}}{\text{molecular mass}}\,(100) = \dfrac{16.0}{18.0} = (100) = 88.9\%$

Notice that the sum of the percent of hydrogen and the percent of oxygen is equal to 100%. Considering rounding in the calculation, the sum of the percentages will always equal 100%. Thus, if all but one of the percentages of composition is known, it is often convenient to merely subtract from 100% to obtain the final value. This method is illustrated in the following example.

Example 2.2: Calculate the percentage composition of glucose, $C_6H_{12}O_6$.

Step 1: Determine the molecular mass of $C_6H_{12}O_6$.

$$\text{carbon} = 6\ C = (6)\ (12.0) = 72.0\ \text{amu}$$
$$\text{hydrogen} = 12\ H = (12)\ (1.0) = 12.0\ \text{amu}$$
$$\underline{\text{oxygen} = 6\ O = (6)\ (16.0) = 96.0\ \text{amu}}$$
$$\text{molecular mass } C_6H_{12}O_6 = 180\ \text{amu}$$

Step 2: Determine the weight percentages.

a. $\% C = \dfrac{\text{mass C}}{\text{molecular mass}}(100) = \dfrac{72.0}{180}(100) = 40.0\%$

b. $\% H = \dfrac{\text{mass H}}{\text{molecular mass}}(100) = \dfrac{12.0}{180}(100) = 6.67\%$

c. $\% O = \dfrac{\text{mass O}}{\text{molecular mass}}(100) = \dfrac{96.0}{180.0}(100) = 53.3\%$

OR:
$$\% C + \% H + \% O = 100\%$$
$$40.0\% + 6.7\% + \% O = 100.0\%$$
$$\% O = 100.0\%$$
$$\% O = 53.3\%$$

By using either method, the percentage for oxygen is the same. This method is especially helpful when there are but two elements in a molecule and the percentage of the second element can be found by subtracting the first percentage from the total or 100%.

2.6 Calculating Empirical Formulas

Let us now consider the reverse operation of the previous section. Suppose the number of atoms of each element in one molecule is unknown. However, the percentage of composition is known. Therefore, the *empirical formula* of the molecule can be determined. The empirical or "simplest" formula is a chemical formula which gives the smallest whole number ratio between the various atoms in the molecule. By writing these whole number ratios as subscripts, an empirical number is obtained.

Example 2.3: Calculate the empirical formula of a compound which is 69.6% oxygen and 30.4% nitrogen by mass.

Step 1: Assume a total mass of 100 grams.

$$\text{Therefore:} \quad \underline{\begin{aligned} 69.6\% \text{ oxygen} &= 69.6 \text{ grams O} \\ 30.4\% \text{ nitrogen} &= 30.4 \text{ grams N} \end{aligned}}$$

$$100 \text{ grams of compound}$$

Step 2: Determine the number of atoms of each element present.

a. $\# \text{ atoms O} = \dfrac{\# \text{ grams oxygen}}{16.0 \text{ grams/atom}} = \dfrac{69.6 \text{ grams}}{16.0 \text{ grams/atom}} = 4.35 \text{ atoms}$

b. $\# \text{ atoms N} = \dfrac{\# \text{ grams nitrogen}}{14.0 \text{ grams/atom}} = \dfrac{30.4 \text{ grams}}{14.0 \text{ grams/atom}} = 2.17 \text{ atoms}$

Step 3: Obtain smallest whole number ratio.

Solution: To obtain the smallest whole number ratio divide the number of atoms of each species by the smallest number of atoms of any element present.

$$N_{2.17}O_{4.35} = N_{\frac{2.17}{2.17}}O_{\frac{4.35}{2.17}} = NO_2 = \text{empirical formula}$$

Example 2.4: Determine the empirical formula of a substance which contains 27.3% carbon and 72.7% oxygen by mass.

Step 1: Assume a total mass of 100 grams.

$$\underline{\begin{aligned} 27.3\% \text{ C} &= 27.3 \text{ grams C} \\ 72.7\% \text{ O} &= 72.7 \text{ grams O} \end{aligned}}$$

$$100 \text{ grams of compound}$$

Step 2: Determine the number of atoms of each element.

a. $\# \text{ atoms C} = \dfrac{\# \text{ grams carbon}}{12.0 \text{ grams/atom}} = \dfrac{27.3 \text{ grams}}{12.0 \text{ grams/atom}} = 2.28 \text{ atoms}$

b. $\# \text{ atoms O} = \dfrac{\# \text{ grams oxygen}}{16.0 \text{ grams/atom}} = \dfrac{72.7 \text{ grams}}{16.0 \text{ grams/atom}} = 4.56 \text{ atoms}$

Step 3: Obtain smallest whole number ratio.

$$C_{2.28}O_{4.56} = C_{\frac{2.28}{2.28}}O_{\frac{4.56}{2.28}} = CO_2 = \text{empirical formula}$$

Calculations relating to percentage composition and empirical formulas are also done using grams as the unit of mass rather than amu. When this is done, the terms molecule and atom are replaced by the terms *mole* and *gram-atom*.

A *mole* is an abbreviation or term used for "gram-molecular weight." By definition, 1 mole of a substance is 6.02×10^{23} molecules of that substance. From the previous example, one mole of carbon dioxide would contain 6×10^{23} molecules of CO_2 and weigh 44 grams. To calculate the number of moles of a substance present, divide the number of grams by the gram-molecular weight:

$$\# \text{ moles} = \frac{\# \text{ grams}}{\text{GMW}}$$

GMW = gram-molecular weight

A *gram-atom* is the term used for "gram-atomic weight." By definition, 1 gram-atom of a substance is 6.02×10^{23} atoms of that substance. For example, 1 gram-atom of sulfur would contain 6.02×10^{23} atoms of sulfur and weigh 32 grams. The number of gram-atoms of a substance is calculated in much the same manner as the number of moles. That is, the number of gram-atoms is equal to the number of grams divided by the gram-atomic weight of the material.

$$\# \text{ gram-atoms} = \frac{\# \text{ grams}}{\text{GAW}} \qquad \text{GAW} = \frac{\text{the mass of a}}{\text{gram-atomic weight}}$$

2.7 Determing Molecular Formulas

By extending the calculation for the empirical formula one additional step, the *molecular formula* or the formula which gives the actual number of atoms of each element in a molecule and not just a whole number ratio, can be obtained. The key to the determination of the molecular formula is the molecular mass or weight. In order to determine the molecular formula from the empirical formula, the molecular mass must be given; or it must be capable of being calculated from additional experimental data.

Example 2.5: A compound is found to have an empirical formula of NO_2. The molecular mass of the compound is known to be 92 amu's/molecule. Calculate the molecular formula.

Step 1: Determine the molecular mass of the empirical formula.

$$NO_2 = (14) + (2)(16) = 46 \text{ amu/molecule.}$$

Step 2: Divide the molecular mass by the mass of the empirical formula.

$$\frac{\text{molecular mass}}{\text{empirical mass}} = \frac{92 \text{ amu/molecule}}{46 \text{ amu/molecule}} = 2$$

Step 3: Multiply all subscripts in the empirical formula by the coefficient obtained in Step 2.

$$(NO_2)_2 = N_2O_4 = \text{Molecular Formula}$$

Example 2.6: A compound is known to have an empirical formula of CH_2O and a molecular mass of 330 amu/molecule. Determine the molecular formula.

Step 1: Mass of the Empirical formula.

$$CH_2O = (12) + (2)(1.0) + (16) = 30 \text{ amu/molecule}$$

Step 2: Ratio of the masses.

$$\frac{\text{molecular mass}}{\text{empirical mass}} = \frac{330 \text{ amu/molecule}}{30 \text{ amu/molecule}} = 11$$

Step 3: Multiply the subscripts.

$$(CH_2O)_{11} = C_{11}H_{22}O_{11} = \text{Molecular Formula}$$

2.8 Writing Chemical Equations

To this point, we have not considered the chemical changes which various molecules may undergo. By definition, a *chemical reaction* is the process by which given substances change chemical properties in such a manner as to create new products which have different physical and chemical properties than the initial species. A *chemical equation* is the shorthand method used to illustrate a chemical reaction.

In a chemical equation, the original species are called reactants and the substances formed are called products. All chemical equations should be balanced. That is, the total number of atoms of each species in the reactants must be equal to the total number of atoms of each species in the products. The *total* number of atoms of each species is what is important, as in most cases, the arrangement of the atoms will be different in the products than in the reactants.

Chemical equations are balanced by putting the proper coefficient before the necessary molecules. **Never change a subscript to balance an equation.** Consider the following chemical reaction:

Step 1: $H_2 + O_2 \rightarrow H_2O$

This equation is read, "One molecule of hydrogen plus one molecule of oxygen yields one molecule of water." This is an unbalanced or skeletal equation.

By inspection, there are two hydrogen atoms and two oxygen atoms on the left-side of the equation and two hydrogen atoms and *one* oxygen atom on the right-side. Therefore, one additional oxygen atom is needed on the right-side. Thus, a coefficient of 2 is inserted before the water molecule.

Step 2: $H_2 + O_2 \rightarrow 2H_2O$

Now there are two hydrogen atoms and two oxygen atoms on the left-side and *four* hydrogen atoms and two oxygen atoms on the right-side. To finish balancing the equation, put a coefficient of 2 before the hydrogen on the left-side of the equation.

Step 3: $2H_2 + O_2 \rightarrow 2H_2O$

The equation is read, "Two molecules of hydrogen plus one molecule of oxygen yields two molecules of water." This is a balanced molecular equation for the reaction between hydrogen and oxygen to form water. Only practice will allow you to balance these simple equations by inspection. You will never learn the method of balancing by watching someone else do it.

Glossary

Atomic Mass Unit—(amu) A unit of mass exactly equal to one-twelfth the mass of carbon-12 atom.

Atomic Mass (Atomic Weight)—The weighted average mass of all the known isotopes of a given element based on the mass of the carbon-12 isotope.

Chemical Equation—The symbolic representation of a chemical reaction. A chemical equation can be written either in words or in chemical symbols.

Chemical Formula—A proper combination of symbols and subscripts representing the composition of a pure substance.

Electron—A subatomic particle found about the nucleus of an atom. An electron has a negative charge.

Empirical Formula—The experimentally determined formula or the chemical formula which illustrates the relative number of atoms of the component elements in a molecule.

Molecular Formula—The true formula or the chemical formula which illustrates the exact number of atoms of the component elements in a molecule.

Molecular Mass (Molecular Weight)—The mass of one molecule of an element based on the carbon-12 atom. The molecular mass is the sum of the atomic masses of the atoms which compose the molecule.

Neutron—A subatomic particle found in the nucleus of an atom. A neutron has no charge.

Proton—A subatomic particle found in the nucleus of an atom. A proton has a positive charge equal in magnitude to the electron.

Symbol—The representation given each element. Each element has a unique symbol represented by one or two letters.

Exercises

1. When an element has two letters in its chemical symbol, explain why the second letter must be written as a lower-case letter.

2. Write chemical symbols for the following elements.

   a. Chlorine
   b. Sodium
   c. Gold
   d. Argon
   e. Silver

3. Calculate the mass of the following molecules. Express your answer in atomic mass units (amu).

   a. HCl
   b. NaF
   c. ZnO
   d. $PbCl_2$
   e. $Ca(OH)_2$

4. For each of the following molecules, list the number of atoms of each element present and the total number of atoms present in the molecule.

   a. $C_6H_{12}O_6$
   b. N_2O_4
   c. H_3PO_4
   d. CH_3COOH
   e. $ZnCl_2$

5. Calculate the percentage composition for the following substances.

   a. CaO
   b. H_2O_2
   c. CuS
   d. Cu_2S
   e. $NaNO_3$

6. Through experimentation, it was found that 28 amu of iron will combine with 16 amu of sulfur. Calculate the percentage composition of iron sulfide formed.

7. Calculate the empirical formula of the following molecules given the percentage composition.

   a. C, 38.7%; H, 9.7%; O, 51.6%
   b. C, 40.0%; H, 6.7%; O, 53.3%
   c. C, 52.3%; H, 13.0%; O, 34.7%
   d. Fe, 63.6%; S, 36.4%
   e. Fe, 70%; O, 30.0%

8. Refer to problem 7. Given the following molecular masses, determine the molecular formula for each compound.

   a.   62 amu/molecule   d.   88 amu/molecule

   b.   90 amu/molecule   e.   160 amu/molecule

   c.   138 amu/molecule

9. Through experimentation, it was found that 12.7 grams of copper metal just reacts with 3.2 grams of powdered sulfur. Calculate:

   a.   Percentage composition

   b.   Empirical formula

   c.   Given a gram molecular mass of 159 grams/mole, determine the molecular formula.

10. Balance the following equations:

   a. $Fe + S = FeS$

   b. $CaO + CO_2 = CaCO_3$

   c. $Na + Cl_2 = NaCl$

   d. $H_2O = H_2 + O_2$

   e. $Zn + HCl = ZnCl_2 + H_2$

   f. $NaBr + F_2 = NaF + Br_2$

   g. $KClO_3 = KCl + O_2$

   h. $Al + HCl = AlCl_3 + H_2$

   i. $Na + H_2O = NaOH + H_2$

   j. $Ca(OH)_2 + HCl = CaCl_2 + H_2O$

   k. $Al_2(SO_4)_3 + BaCl_2 = BaSO_4 + AlCl_3$

   l. $Cu(NO_3)_2 + H_3PO_4 = Cu_3(PO_4)_2 + HNO_3$

   m. $Bi_2(CO_3)_3 + CH_3COOH = Bi(CH_3COO)_3 + H_2CO_3$

   n. $(NH_4)_3PO_4 + KOH = K_3PO_4 + NH_4OH$

   o. $Zn + AuCl = ZnCl_2 + Au$

   p. $BaO_2 = BaO + O_2$

   q. $S + O_2 = SO_3$

   r. $Pb(NO_3)_2 + Na_2SO_4 = PbSO_4 + NaNO_3$

CHAPTER 3

Periodic Properties and Atomic Theory

3.1 Historical Development of the Periodic Table

Chemistry, like much of science, has been accumulated since early Greco-Roman times. However, it was not until the 19th century that a sufficient amount of knowledge was obtained to make a statement about the periodic pattern that the elements seemed to follow.

In the 19th century, such men as Johann Doebereiner, a German chemist; John Newlands, an English chemist; Lothar Mayer, a German chemist; and Dmitri Mendeleev, a Russian scientist all postulated the existence of some sort of arrangement of periodicity of the elements.

In 1869, Mayer and Mendeleev working independently came to essentially the same conclusion on the arrangement of the elements. This early arrangement of the elements was according to atomic mass and not according to atomic number as it is today.

The genius of Mendeleev is illustrated in that only fifty-six elements were known when he presented his table and yet, he had the foresight to leave blanks where many of the future elements would fit. One family of elements not listed on the early chart was the noble gas family. Once the noble gases were added and the elements arranged according to atomic number, the early view of the modern periodic table was achieved.

3.2 Periodic Table—Families and Periods

Due to the length of the table and the multitude of properties exhibited in the periodic table, it is convenient to divide the table into sections according to the various chemical properties. One method of grouping is the arrangement of the elements into families or periods.

If the long form of the periodic table is studied, there is a series of horizontal rows and vertical columns. Each horizontal row forms what is called a *period* of elements. Each vertical column forms what is called a *family* or *group* of elements. The long form of the periodic table is found in Figure 3.1.

The first period contains but two elements; hydrogen and helium. The second period contains eight elements ranging from lithium to neon. The third period also contains eight elements ranging from sodium to argon. The fourth period is much longer and contains eighteen elements from potassium to krypton. **This is no accident!** Remember this relationship and the numbers 2, 8, 8, 18 as you will soon see how this relates to the electron structure, and thus the chemical properties, of the elements.

As you proceed from left to right across a period, the chemical properties vary greatly. This is due to the electron structure. For example, consider the fourth period. Potassium is a very reactive substance that will burn in air. On the other hand, krypton is essentially unreactive and considered a noble gas. Thus, within a period, the chemical properties have a great deal of variation from element to element.

Now consider the vertical orientation. All the elements in a family have similar electron configurations and therefore similar chemical properties. For example, family or group IA is referred to as the alkali-earth metals and all the elements in this group are very reactive. As we shall see later, they all have a similar electron structure. Another family is the halogens or group VIIA. The members of this group include fluorine, chlorine, bromine, and iodine. Again, these elements all have similar chemical properties; are all electronegative, and their atoms are found in the form of diatomic molecules.

3.3 The Noble Gases

As mentioned in Chapter 2, probably the most important family as far as electron structure is concerned is the noble gases. Members of this family are helium, neon, argon, krypton, xenon, and radon. These elements are the pivot points for each period of elements. Each noble gas has a stable electron structure and is very

Figure 3.1 Periodic Chart With Atomic Masses

	IA	IIA	IIIB	IVB	VB	VIB	VIIB	VIII	VIII	VIII	IB	IIB	IIIA	IVA	VA	VIA	VIIA	0
1	1 H 1.008																	2 He 4.003
2	3 Li 6.939	4 Be 9.012											5 B 10.81	6 C 12.01	7 N 14.01	8 O 16.00	1 H 1.008	10 Ne 20.18
3	11 Na 23.00	12 Mg 24.31											13 Al 26.98	14 Si 28.09	15 P 30.97	16 S 32.06	17 Cl 35.45	18 Ar 39.94
4	19 K 39.10	20 Ca 40.08	21 Sc 44.96	22 Ti 47.90	23 V 50.94	24 Cr 52.00	25 Mn 54.94	26 Fe 55.85	27 Co 58.93	28 Ni 58.71	29 Cu 63.54	30 Zn 65.37	31 Ga 69.72	32 Ge 72.59	33 As 74.92	34 Se 78.96	35 Br 79.91	36 Kr 83.80
5	37 Rb 85.47	38 Sr 87.62	39 Y 88.90	40 Zr 91.22	41 Nb 92.91	42 Mo 95.94	43 Tc (97)	44 Ru 101.1	45 Rh 102.9	46 Pd 106.4	47 Ag 107.9	48 Cd 112.4	49 In 114.8	50 Sn 118.7	51 Sb 121.8	52 Te 127.6	53 I 126.9	54 Xe 131.3
6	55 Cs 132.9	56 Ba 137.3	57 La 138.9	72 Hf 178.5	73 Ta 180.9	74 W 183.8	75 Re 186.2	76 Os 190.2	77 Ir 192.2	78 Pt 195.1	79 Au 197.0	80 Hg 200.6	81 Tl 204.4	82 Pb 207.2	83 Bi 209.0	84 Po (209)	85 At (210)	86 Rn (222)
7	87 Fr (223)	88 Ra (226)	89 Ac (227)															

58 Ce 140.1	59 Pr 140.9	60 Nd 144.2	61 Pm (145)	62 Sm 150.4	63 Eu 152.0	64 Gd 157.2	65 Tb 158.9	66 Dy 162.5	67 Ho 164.9	68 Er 167.3	69 Tm 168.9	70 Yb 173.0	71 Lu 175.0
90 Th 232.0	91 Pa (231)	92 U 238.0	93 Np (237)	94 Pu (244)	95 Am (243)	96 Cm (247)	97 Bk (247)	98 Cf (251)	99 Es (254)	100 Fm (253)	101 Md (256)	102 No (253)	103 Lw (257)

36

TABLE 3.1
ELECTRON STRUCTURE OF THE NOBLE GASES

Element	Number of Electrons	Difference in the number of Electrons between each element
He	2	
Ne	10	8
Ar	18	8
Kr	36	18
Xe	54	18
Rn	86	32

unreactive. The number of electrons for each element is tabulated in Table 3.1. Be able to recall this relationship.

3.4 Metals and Nonmetals

Another general method of grouping the elements is according to one of their chemical or physical properties. One such method is to call them either metals or nonmetals. A *metal* is defined as any element having properties such as ductility, malleability, and conductivity. A *nonmetal* is defined as any element having properties such as low luster, brittleness, and being a nonconductor. It must be noted that the transition between metals and nonmetals is not a black and white distinction but rather a gradual change. For convenience of study and discussion, there is a division line (broken line in Figure 3.1) drawn between the metals and nonmetals. This line may be used for discussion purposes but remember that the change is gradual and a chemist must be extremely careful in making judgements solely on this basis.

3.5 Transition Elements

As mentioned in the previous section, the change between metals and nonmetals is gradual. The elements in Groups IIIB-VIIB and Group VIII are known as *transition elements* or elements which form a multitude of compounds and show a variety of electron configurations. All transition elements are considered metals but they do show variation in properties from the very active metals to the nonmetals.

For the beginning student, the transition elements are fairly complex as they illustrate variations in electron structure which in turn changes the chemical bonding and consequently the chemical compounds formed. Keeping this in mind will be of aid when studying the electron configurations of the transition elements in the next chapter.

3.6 Rutherford and Atomic Structure

There have been many experiments performed in attempting to determine the actual structure of atoms. One of the earlier attempts which provided a great deal of pertinent information was the work of Lord Rutherford of England in 1911. Rutherford's conclusion was that the protons were gathered together in a very small region called the nucleus at the center of the atom.

From the work of other scientists, the charge and mass relationships of the atom were determined. Using this additional information, Rutherford gathered all available data and performed his experiment. The actual experiment involved the bombardment of a very thin piece of gold foil with alpha particles (helium nuclei) which are positively charged. Behind the gold foil was centered a piece of photographic film to detect the alpha particles after they had passed through the gold foil. See Figure 3.2.

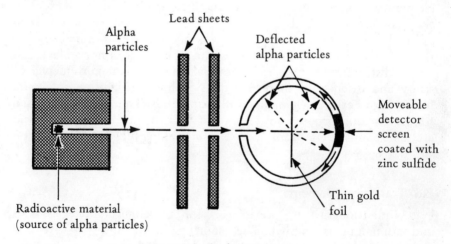

Figure 3.2 Rutherfords' Experiment

The results of the experiment indicated that most of the alpha particles passed directly through the gold foil. This indicated that the gold foil was mostly just "space." About one in every 100,000 alpha particles was deflected at an appreciable angle. This indicated there was a repulsive force caused by regions of positive charge in the gold foil.

From these two observations came the visualizing of the atom as an area of mostly empty space but within this space was located a dense, very small, positively charged nucleus. Later experiments were to indicate that over 99.9% of the mass of an atom is located in the nucleus. In chapter four we will see how chemists were to take this information and develop the present day concept of the atom.

Glossary

Family (Group)—A given column of elements of the periodic table with similar chemical properties. An example is the noble gases.

Metal—Any element which shows such properties as conductivity, ductility, malleability, and luster.

Noble Gases—A family of relatively unreactive gases. Members of the family are He, Ne, Ar, Kr, Xe, and Rn.

Nonmetal—Any element which does not exhibit the metallic properties.

Nucleus—The center of an atom. The nucleus is composed of protons and neutrons thus having a net positive charge.

Period—A given row of the periodic chart. There are presently seven periods in the periodic table.

Periodic Chart—A listing or arrangement of the elements according to increasing atomic number.

Transition Elements—The elements in the 3rd—8th periods of the periodic chart. They are all metals.

Exercises

1. Explain why Rutherford was able to conclude that atoms had a positively charged nucleus.

2. Arrange the following elements in order of increasing electronegativity. That is, list the most electronegative element last.

   a. lithium   d. oxygen

   b. boron   e. nitrogen

   c. fluorine

3. List four elements which are members of a family of elements called the halogens (Column VIIA).

4. Classify each of the following as either a metal or nonmetal.

   a. Mn   d. Ba

   b. As   e. S

   c. Te

5. Which of the following are diatomic gases under normal conditions?

a. neon

d. sodium

b. fluorine

e. iodine

c. oxygen

6. What would have happened in Rutherford's experiment if he had used beta particles (electrons) in place of alpha particles for his source of radiation?

7. Why isn't astatine discussed along with the other halogens as being a diatomic molecule? What unique chemical properties does astatine have?

8. Why isn't radon discussed along with the other noble gases?

9. List the members of the Lanthanide series. What properties do they have in common?

10. List the members of the Actinide series. Which of these members are considered artificial or man-made elements?

CHAPTER 4

The Electron
Arrangement of Atoms

4.1 Introduction to Electron Arrangement

In the previous chapters, some of the early concepts of atoms were discussed and a brief description given of the various particles which compose atoms.

From the periodic table, it is apparent that there is a definite arrangement of the atoms with respect to their chemical behavior. Much of this behavior depends on the arrangement of the electrons about the nucleus. The relationship between the electron and energy levels has been of great interest to the chemist since the early history of science.

It must be emphasized that the concepts presented in the next chapters on structure and bonding are very abstract and theoretical. Chemists, themselves, do not agree completely on all the theories presented. However, it is felt that an accurate, composite picture of an atom is now visualized by combining the many theories of structure and bonding together.

Through the use of instrumentation, evidence is continually being gathered to substantiate the theories of atomic structure. The study of the spectra of elements such as hydrogen and helium has contributed invaluable information. However, remember no one has actually seen an atom so the theories presented are those which modern chemists feel best describe what they envision the arrangement of atoms to be; supported by a great deal of experimentation and calculation.

4.2 Quantum Numbers

One of the early, quantum mechanical concepts of an atom was the Bohr model of the atom as conceived by Niels Bohr, a Danish scientist in the early twentieth-century. Bohr's Theory of the atom had four basic postulates.

1. The electron in an atom has only certain definite stationary states of motion allowed to it; each of these stationary states has a definite fixed energy.
2. When an electron is in one of these stationary states, it does not radiate; but when changing from a high energy state to a state of lower energy, the electron emits a quantum of radiation whose energy is equal to the difference in the energy of the two states.
3. The states of allowed electron movement are those in which the angular momentum of the electron is an integral multiple of $h/2\pi$.
4. In any of the allowed energy states, the electron moves in a circular orbit about the nucleus.

From these postulates, it can be seen that the Bohr model of the atom had the electrons revolving about a small fixed nucleus. This idea is often called the planetary concept of an atom as it has much the same orientation as our solar system.

As is the case with many theories, time has caused some adjustment and change. In place of the circular orbits, a three-dimensional sphere is now theorized to be more appropriate. The idea of quantized energy is still a fundamental concept. Quantizing of energy means that energy is released in little bundles or packages. That is, energy is released in steps and not at a continual flow. Thus, about the nucleus of an atom there are a series of distinct energy levels as allowed by the mathematics of Bohr's theory. When an electron moves from a higher energy level to a lower energy level, energy is released.

Through improved instrumentation, measurements of the radiation emitted by excited atoms revealed that the energies of electrons within a given energy level may vary. Therefore, it was postulated that within a main energy level there must be energy sub-levels to account for these variations.

By taking the above information and a great deal of mathematics into account, scientists have developed a description of atomic structure that represents the electron in a given atom in terms of *quantum mechanics.*

A portion of quantum theory describes the electrons in a sub-level as in constant motion, with this motion confined to a region of space called an *orbital*. There are four types of orbitals as designated by the letters *s, p, d,* and *f* in the *ground state* or lowest energy state for atoms. The reason for the letters comes from their origin in spectroscopy where the letters are used to label different series of spectral lines as "sharp, principal, diffuse, and fundamental." Figure 4.1 illustrates the shapes of the various orbitals.

An electron can be anywhere within an orbital at a given time with certain portions of the orbital favored over others. No more than two electrons may occupy each orbital and the two electrons in the orbital have a spin which in all cases must be opposite to each other.

It is obvious by this time that all the information dealing with quantum theory is involved and somewhat difficult to remember in the present format. Fortunately, all the above information can be summarized with the use of *quantum numbers*. There are four quantum numbers as summarized in Table 4.1.

TABLE 4.1
QUANTUM NUMBERS DESCRIBED

Quantum Number	Name	Values or Designation	Information
1	Principal	1, 2, 3, . .	Describes which main energy level the electron is in.
2	Secondary or Angular Momentum	*s, p, d, f,* . .	Indicates which sub-level the electron is in.
3	Magnetic	Dependent on the sub-level	Explains which orbital the electron is in within a given sub-level.
4	Spin	$+\frac{1}{2}, -\frac{1}{2}$	Indicates the direction the electron is rotating within an orbital.

It must be noted that the number of sub-levels and consequently the number of orbitals and electrons is dependent upon the energy level occupied.

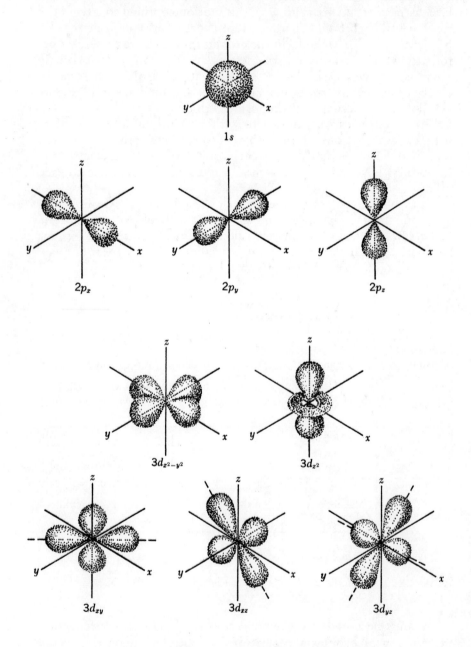

Figure 4.1. A Three-Dimensional View of the *s*, *p*, and *d* Orbitals.

There is but one sub-level of the first energy level with one orbital (s), in that sub-level. There are two sub-levels in the second energy level with four orbitals (1s and 3 p's), in that sub-level. In the third level there are three sub-levels with nine orbitals (1s, 3 p's, and 5 d's) in that sub-level. To summarize this information, see Table 4.2.

TABLE 4.2
RELATIONSHIP BETWEEN THE NUMBER OF ELECTRONS,
ORBITALS AND THE ENERGY LEVEL

Energy Level

	1	2	3	4
Number of Orbitals	1	4	9	16
Type of Orbitals	s 1	s p 1 3	s p d 1 3 5	s p d f 1 3 5 7
Max. # e^-/Orbital	2 ,	2 6	2 6 10	2 6 10 14
Total # of e^-	2	8	18	32

4.3 Writing Electron Configurations

To this point in the discussion of electronic structure, many general statements have been made. In addition, a great deal of abstract theory has been presented in various forms. One way to comprehend the topic of electron arrangement is to apply the previous information to the elements of the periodic chart and observe the pattern which develops.

As mentioned in section 4.2, there is a definite order in which the electrons fill the various energy levels about the nucleus. Figure 4.2 illustrates the order of filling as well as the relationship between energy levels and orbitals. In studying Figure 4.2, notice that the lowest energy level ($n = 1$), is at the bottom of the figure. In writing electron configurations for the various atoms, *the lowest energy level is always filled first.* Then the second level is filled and so on up the chart. Remember that each orbital, no matter what type, can hold but two electrons. The principle that electrons fill from lowest available energy level first is known as the *Aufbau Principle.*

By use of the above diagram, the schematic representation as well as the electron configuration for any atom may be written. Consider the first element on the periodic chart, hydrogen. Hydrogen has

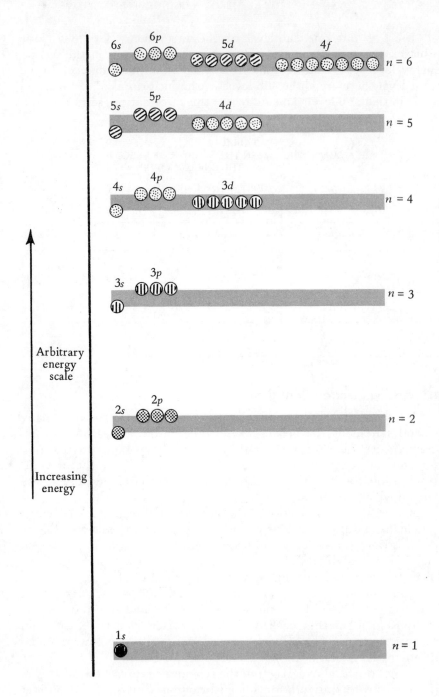

Figure 4.2. Energy Level Diagram.

46

atomic number 1 and therefore, has one electron. By use of the Aufbau Principle, the single electron in the hydrogen atom would be in the first energy level and thus an *s* orbital. See Figure 4.3. Study

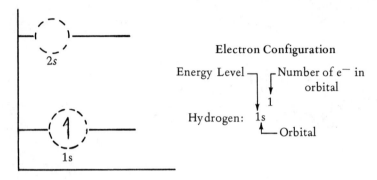

Figure 4.3. Electron Configuration for Hydrogen Atom.

the electron configuration for hydrogen. The large coefficient 1 means the electron is in the first energy level. The "*s*" means the electron is in an "*s*" orbital and the exponent 1 means there is but one electron in the *s* orbital.

Now study Figure 4.4 for the electron configuration and schematic diagram for helium; atomic number 2. Notice helium has one

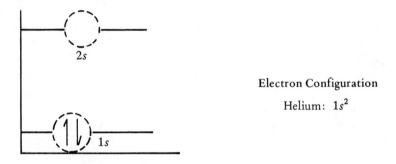

Figure 4.4. Electron Configuration of the Helium Atom.

additional electron which according to the Aufbau Principle would go to the lowest possible orbital which is still the 1*s* orbital because each orbital may hold a maximum of two electrons. The electron configuration for helium is $1s^2$ indicating the first energy level, *s* orbital with 2 electrons.

Remember the two electrons in a given orbital must have opposite spins. This is because no two electrons in the same atom can have the same set of quantum numbers. This statement or principle of quantum theory is known as the *Pauli Exclusion Principle.*

Next, consider the configuration for carbon. Study Figure 4.5 for the schematic diagram and electron configuration for carbon. Carbon

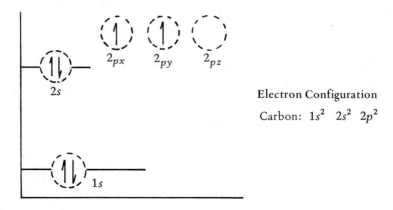

Electron Configuration

Carbon: $1s^2 \quad 2s^2 \quad 2p^2$

Figure 4.5. Electron Configuration for the Carbon Atom.

has two electrons in the $1s$ orbital, two electrons in the next higher energy orbital which is the $2s$ and then 2 electrons in the $2p$ orbital. Notice the $2p$ orbitals. There is one electron in the P_x orbital and one electron in the p_y orbital. It is extremely important to remember this pattern of filling as it is the key in the determination of chemical bonds. It should also be noted that the x, y, and z designation of the orbitals is arbitrary with respect to the orbitals' orientation in space. When orbitals have the same energy as in the case with the three $2p$ orbitals, the electrons will fill each one individually as long as possible and then will return and fill a second electron in each of the orbitals. This principle of filling is often referred to as *Hund's Rule.*

Table 4.3 lists the schematic diagram and electron configurations for the first 36 elements. The symbol (Ar) means the argon configuration is written followed by the configuration listed. This is a common shorthand method of writing electron configurations.

From Table 4.3 study the electron configurations for helium, neon, argon, and krypton. Each is a member of the noble gases. Each element has a completed energy level. This is an extremely important property. The noble gases are very stable and are essentially unreac-

TABLE 4.3
ELECTRON CONFIGURATIONS FOR THE FIRST 36 ELEMENTS

ELEMENT	ELECTRON CONFIGURATION	SCHEMATIC REPRESENTATION

ELEMENT	ELECTRON CONFIGURATION	$1s$	$2s$	$2p$	$3s$	$3p$
H	$1s^1$	↑				
He	$1s^2$	↑↓				
Li	$1s^2 2s^1$	↑↓	↑	○○○		
Be	$1s^2 2s^2$	↑↓	↑↓	○○○		
B	$1s^2 2s^2 2p^1$	↑↓	↑↓	↑ ○ ○		
C	$1s^2 2s^2 2p^2$	↑↓	↑↓	↑ ↑ ○		
N	$1s^2 2s^2 2p^3$	↑↓	↑↓	↑ ↑ ↑		
O	$1s^2 2s^2 2p^4$	↑↓	↑↓	↑↓ ↑ ↑		
F	$1s^2 2s^2 2p^5$	↑↓	↑↓	↑↓ ↑↓ ↑		
Ne	$1s^2 2s^2 2p^6$	↑↓	↑↓	↑↓ ↑↓ ↑↓		
Na	$1s^2 2s^2 2p^6 3s^1$	↑↓	↑↓	↑↓ ↑↓ ↑↓	↑	○○○
Mg	$1s^2 2s^2 2p^6 3s^2$	↑↓	↑↓	↑↓ ↑↓ ↑↓	↑↓	○○○
Al	$1s^2 2s^2 2p^6 3s^2 3p^1$	↑↓	↑↓	↑↓ ↑↓ ↑↓	↑↓	↑ ○ ○
Si	$1s^2 2s^2 2p^6 3s^2 3p^2$	↑↓	↑↓	↑↓ ↑↓ ↑↓	↑↓	↑ ↑ ○
P	$1s^2 2s^2 2p^6 3s^2 3p^3$	↑↓	↑↓	↑↓ ↑↓ ↑↓	↑↓	↑ ↑ ↑
S	$1s^2 2s^2 2p^6 3s^2 3p^4$	↑↓	↑↓	↑↓ ↑↓ ↑↓	↑↓	↑↓ ↑ ↑
Cl	$1s^2 2s^2 2p^6 3s^2 3p^5$	↑↓	↑↓	↑↓ ↑↓ ↑↓	↑↓	↑↓ ↑↓ ↑
Ar	$1s^2 2s^2 2p^6 3s^2 3p^6$	↑↓	↑↓	↑↓ ↑↓ ↑↓	↑↓	↑↓ ↑↓ ↑↓

ELEMENT	CONFIGURATION	$3s$	$3p$	$3d$	$4s$	$4p$
K	$(Ar)4s^1$	↑↓	↑↓ ↑↓ ↑↓	○○○○○	↑	○○○
Ca	$(Ar)4s^2$	↑↓	↑↓ ↑↓ ↑↓	○○○○○	↑↓	○○○
Sc	$(Ar)3d^1 4s^2$	↑↓	↑↓ ↑↓ ↑↓	↑ ○○○○	↑↓	○○○
Ti	$(Ar)3d^2 4s^2$	↑↓	↑↓ ↑↓ ↑↓	↑ ↑ ○○○	↑↓	○○○
V	$(Ar)3d^3 4s^2$	↑↓	↑↓ ↑↓ ↑↓	↑ ↑ ↑ ○○	↑↓	○○○
Cr	$(Ar)3d^5 4s^1$	↑↓	↑↓ ↑↓ ↑↓	↑ ↑ ↑ ↑ ↑	↑	○○○
Mn	$(Ar)3d^5 4s^2$	↑↓	↑↓ ↑↓ ↑↓	↑ ↑ ↑ ↑ ↑	↑↓	○○○
Fe	$(Ar)3d^6 4s^2$	↑↓	↑↓ ↑↓ ↑↓	↑↓ ↑ ↑ ↑ ↑	↑↓	○○○
Co	$(Ar)3d^7 4s^2$	↑↓	↑↓ ↑↓ ↑↓	↑↓ ↑↓ ↑ ↑ ↑	↑↓	○○○
Ni	$(Ar)3d^8 4s^2$	↑↓	↑↓ ↑↓ ↑↓	↑↓ ↑↓ ↑↓ ↑ ↑	↑↓	○○○
Cu	$(Ar)3d^{10} 4s^1$	↑↓	↑↓ ↑↓ ↑↓	↑↓ ↑↓ ↑↓ ↑↓ ↑↓	↑	○○○
Zn	$(Ar)3d^{10} 4s^2$	↑↓	↑↓ ↑↓ ↑↓	↑↓ ↑↓ ↑↓ ↑↓ ↑↓	↑↓	○○○
Ga	$(Ar)3d^{10} 4s^2 4p^1$	↑↓	↑↓ ↑↓ ↑↓	↑↓ ↑↓ ↑↓ ↑↓ ↑↓	↑↓	↑ ○ ○
Ge	$(Ar)3d^{10} 4s^2 4p^2$	↑↓	↑↓ ↑↓ ↑↓	↑↓ ↑↓ ↑↓ ↑↓ ↑↓	↑↓	↑ ↑ ○
As	$(Ar)3d^{10} 4s^2 4p^3$	↑↓	↑↓ ↑↓ ↑↓	↑↓ ↑↓ ↑↓ ↑↓ ↑↓	↑↓	↑ ↑ ↑
Se	$(Ar)3d^{10} 4s^2 4p^4$	↑↓	↑↓ ↑↓ ↑↓	↑↓ ↑↓ ↑↓ ↑↓ ↑↓	↑↓	↑↓ ↑ ↑
Br	$(Ar)3d^{10} 4s^2 4p^5$	↑↓	↑↓ ↑↓ ↑↓	↑↓ ↑↓ ↑↓ ↑↓ ↑↓	↑↓	↑↓ ↑↓ ↑
Kr	$(Ar)3d^{10} 4s^2 4p^6$	↑↓	↑↓ ↑↓ ↑↓	↑↓ ↑↓ ↑↓ ↑↓ ↑↓	↑↓	↑↓ ↑↓ ↑↓

(Note: $3d$ labeled "diffuse")

tive at normal conditions. A deduction from the properties of noble gases is that the most stable configuration an element can achieve is that of a completed energy level or noble gas configuration. In the next chapter, this will be a very important factor in the discussion of bonding and ionization.

The schematic representation for the first thirty-six elements follows the Aufbau Principle and Hund's Rule in determining the order of filling. There are times when a memory device can be useful in remembering the order of filling of the various subshells. Such a device is shown in Figure 4.6. Simply follow the arrows inserting the proper number of electrons in each subshell starting with the 1s orbital.

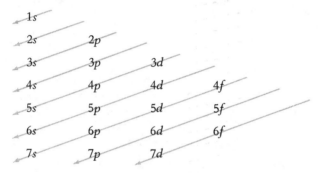

Figure 4.6. A Memory Device for the Filling of Electrons in Subshells.

Unfortunately, the order of filling does not follow this continual pattern throughout the entire periodic chart. From Figure 4.3, it can be noted that the energy levels become closer and closer together as the energy level number increases. That is, energy levels 4 and 5 are much closer together than are energy levels 1 and 2. As a result, in many ~~electron~~ atoms there is a tendency of electrons to vary in the order of filling as the energy differences become smaller and smaller.

In the transition elements there becomes a stability of half-filled orbitals in addition to completely filled orbitals. This variation involves many aspects other than just energy or order of filling. At this point in the study of structure and bonding, it is sufficient to work with the lower numbered elements and accept the configuration of the transition elements. Table 4.4 lists the electron configurations for all the elements. By referring to this tabulation, the configuration for any element in the ground state or lowest energy state may be obtained.

TABLE 4.4
ELECTRON CONFIGURATIONS FOR THE ELEMENTS

Atomic number	Element	1 s	2 s p	3 s p d	4 s p d f	5 s p d f	6 s p d	7 s
1	H	1						
2	He	2						
3	Li	2	1					
4	Be	2	2					
5	B	2	2 1					
6	C	2	2 2					
7	N	2	2 3					
8	O	2	2 4					
9	F	2	2 5					
10	Ne	2	2 6					
11	Na	2	2 6	1				
12	Mg	2	2 6	2				
13	Al	2	2 6	2 1				
14	Si	2	2 6	2 2				
15	P	2	2 6	2 3				
16	S	2	2 6	2 4				
17	Cl	2	2 6	2 5				
18	Ar	2	2 6	2 6				
19	K	2	2 6	2 6	1			
20	Ca	2	2 6	2 6	2			
21	Sc	2	2 6	2 6 1	2			
22	Ti	2	2 6	2 6 2	2			
23	V	2	2 6	2 6 3	2			
24	Cr	2	2 6	2 6 5	1			
25	Mn	2	2 6	2 6 5	2			
26	Fe	2	2 6	2 6 6	2			
27	Co	2	2 6	2 6 7	2			
28	Ni	2	2 6	2 6 8	2			
29	Cu	2	2 6	2 6 10	1			
30	Zn	2	2 6	2 6 10	2			
31	Ga	2	2 6	2 6 10	2 1			
32	Ge	2	2 6	2 6 10	2 2			
33	As	2	2 6	2 6 10	2 3			
34	Se	2	2 6	2 6 10	2 4			
35	Br	2	2 6	2 6 10	2 5			
36	Kr	2	2 6	2 6 10	2 6			
37	Rb	2	2 6	2 6 10	2 6	1		
38	Sr	2	2 6	2 6 10	2 6	2		
39	Y	2	2 6	2 6 10	2 6 1	2		
40	Zr	2	2 6	2 6 10	2 6 2	2		
41	Nb	2	2 6	2 6 10	2 6 4	1		
42	Mo	2	2 6	2 6 10	2 6 5	1		
43	Tc	2	2 6	2 6 10	2 6 6	1		
44	Ru	2	2 6	2 6 10	2 6 7	1		
45	Rh	2	2 6	2 6 10	2 6 8	1		
46	Pd	2	2 6	2 6 10	2 6 10			
47	Ag	2	2 6	2 6 10	2 6 10	1		
48	Cd	2	2 6	2 6 10	2 6 10	2		
49	In	2	2 6	2 6 10	2 6 10	2 1		
50	Sn	2	2 6	2 6 10	2 6 10	2 2		
51	Sb	2	2 6	2 6 10	2 6 10	2 3		
52	Te	2	2 6	2 6 10	2 6 10	2 4		

TABLE 4.4. (Con't.)

Atomic number	Element	1 s	2 s p	3 s p d	4 s p d f	5 s p d f	6 s p d	7 s
53	I	2	2 6	2 6 10	2 6 10	2 5		
54	Xe	2	2 6	2 6 10	2 6 10	2 6		
55	Cs	2	2 6	2 6 10	2 6 10	2 6	1	
56	Ba	2	2 6	2 6 10	2 6 10	2 6	2	
57	La	2	2 6	2 6 10	2 6 10	2 6 1	2	
58	Ce	2	2 6	2 6 10	2 6 10 1	2 6 1	2	
59	Pr	2	2 6	2 6 10	2 6 10 3	2 6	2	
60	Nd	2	2 6	2 6 10	2 6 10 4	2 6	2	
61	Pm	2	2 6	2 6 10	2 6 10 5	2 6	2	
62	Sm	2	2 6	2 6 10	2 6 10 6	2 6	2	
63	Eu	2	2 6	2 6 10	2 6 10 7	2 6	2	
64	Gd	2	2 6	2 6 10	2 6 10 7	2 6 1	2	
65	Tb	2	2 6	2 6 10	2 6 10 9	2 6	2	
66	Dy	2	2 6	2 6 10	2 6 10 10	2 6	2	
67	Ho	2	2 6	2 6 10	2 6 10 11	2 6	2	
68	Er	2	2 6	2 6 10	2 6 10 12	2 6	2	
69	Tm	2	2 6	2 6 10	2 6 10 13	2 6	2	
70	Yb	2	2 6	2 6 10	2 6 10 14	2 6	2	
71	Lu	2	2 6	2 6 10	2 6 10 14	2 6 1	2	
72	Hf	2	2 6	2 6 10	2 6 10 14	2 6 2	2	
73	Ta	2	2 6	2 6 10	2 6 10 14	2 6 3	2	
74	W	2	2 6	2 6 10	2 6 10 14	2 6 4	2	
75	Re	2	2 6	2 6 10	2 6 10 14	2 6 5	2	
76	Os	2	2 6	2 6 10	2 6 10 14	2 6 6	2	
77	Ir	2	2 6	2 6 10	2 6 10 14	2 6 7	2	
78	Pt	2	2 6	2 6 10	2 6 10 14	2 6 9	1	
79	Au	2	2 6	2 6 10	2 6 10 14	2 6 10	1	
80	Hg	2	2 6	2 6 10	2 6 10 14	2 6 10	2	
81	Tl	2	2 6	2 6 10	2 6 10 14	2 6 10	2 1	
82	Pb	2	2 6	2 6 10	2 6 10 14	2 6 10	2 2	
83	Bi	2	2 6	2 6 10	2 6 10 14	2 6 10	2 3	
84	Po	2	2 6	2 6 10	2 6 10 14	2 6 10	2 4	
85	At	2	2 6	2 6 10	2 6 10 14	2 6 10	2 5	
86	Rn	2	2 6	2 6 10	2 6 10 14	2 6 10	2 6	
87	Fr	2	2 6	2 6 10	2 6 10 14	2 6 10	2 6	1
88	Ra	2	2 6	2 6 10	2 6 10 14	2 6 10	2 6	2
89	Ac	2	2 6	2 6 10	2 6 10 14	2 6 10	2 6 1	2
90	Th	2	2 6	2 6 10	2 6 10 14	2 6 10	2 6 2	2
91	Pa	2	2 6	2 6 10	2 6 10 14	2 6 10 2	2 6 1	2
92	U	2	2 6	2 6 10	2 6 10 14	2 6 10 3	2 6 1	2
93	Np	2	2 6	2 6 10	2 6 10 14	2 6 10 4	2 6 1	2
94	Pu	2	2 6	2 6 10	2 6 10 14	2 6 10 6	2 6	2
95	Am	2	2 6	2 6 10	2 6 10 14	2 6 10 7	2 6	2
96	Cm	2	2 6	2 6 10	2 6 10 14	2 6 10 7	2 6 1	2
97	Bk	2	2 6	2 6 10	2 6 10 14	2 6 10 9	2 6	2
98	Cf	2	2 6	2 6 10	2 6 10 14	2 6 10 10	2 6	2
99	Es	2	2 6	2 6 10	2 6 10 14	2 6 10 11	2 6	2
100	Fm	2	2 6	2 6 10	2 6 10 14	2 6 10 12	2 6	2
101	Md	2	2 6	2 6 10	2 6 10 14	2 6 10 13	2 6	2
102	No	2	2 6	2 6 10	2 6 10 14	2 6 10 14	2 6	2
103	Lw	2	2 6	2 6 10	2 6 10 14	2 6 10 14	2 6 1	2

Periodic Chart (elements with electron configurations):

IA	IIA	IIIB	IVB	VB	VIB	VIIB	VIII			IB	IIB	IIIA	IVA	VA	VIA	VIIA	0
1 H $1s^1$																	2 He $1s^2$
3 Li $2s^1$	4 Be $2s^2$											5 B $2p^1$	6 C $2p^2$	7 N $2p^3$	8 O $2p^4$	9 F $2p^5$	10 Ne $2p^6$
11 Na $3s^1$	12 Mg $3s^2$											13 Al $3p^1$	14 Si $3p^2$	15 P $3p^3$	16 S $3p^4$	17 Cl $3p^5$	18 Ar $3p^6$
19 K $4s^1$	20 Ca $4s^2$	21 Sc $3d^14s^2$	22 Ti $3d^24s^2$	23 V $3d^34s^2$	24 Cr $3d^54s^1$	25 Mn $3d^54s^2$	26 Fe $3d^64s^2$	27 Co $3d^74s^2$	28 Ni $3d^84s^2$	29 Cu $3d^{10}4s^1$	30 Zn $3d^{10}4s^2$	31 Ga $4p^1$	32 Ge $4p^2$	33 As $4p^3$	34 Se $4p^4$	35 Br $4p^5$	36 Kr $4p^6$
37 Rb $5s^1$	38 Sr $5s^2$	39 Y $4d^15s^2$	40 Zr $4d^25s^2$	41 Nb $4d^45s^1$	42 Mo $4d^55s^1$	43 Tc $4d^55s^2$	44 Ru $4d^75s^1$	45 Rh $4d^85s^1$	46 Pd $4d^{10}$	47 Ag $4d^{10}5s^1$	48 Cd $4d^{10}5s^2$	49 In $5p^1$	50 Sn $5p^2$	51 Sb $5p^3$	52 Te $5p^4$	53 I $5p^5$	54 Xe $5p^6$
55 Cs $6s^1$	56 Ba $6s^2$	57 La° $5d^16s^2$	72 Hf $5d^26s^2$	73 Ta $5d^36s^2$	74 W $5d^46s^2$	75 Re $5d^56s^2$	76 Os $5d^66s^2$	77 Ir $5d^76s^2$	78 Pt $5d^96s^1$	79 Au $5d^{10}6s^1$	80 Hg $5d^{10}6s^2$	81 Tl $6p^1$	82 Pb $6p^2$	83 Bi $6p^3$	84 Po $6p^4$	85 At $6p^5$	86 Rn $6p^6$
87 Fr $7s^1$	88 Ra $7s^2$	89 Ac† $6d^17s^2$															

° Lanthanides

58 Ce $4f^15d^16s^2$	59 Pr $4f^36s^2$	60 Nd $4f^46s^2$	61 Pm $4f^56s^2$	62 Sm $4f^66s^2$	63 Eu $4f^76s^2$	64 Gd $4f^75d^16s^2$	65 Tb $4f^96s^2$	66 Dy $4f^{10}6s^2$	67 Ho $4f^{11}6s^2$	68 Er $4f^{12}6s^2$	69 Tm $4f^{13}6s^2$	70 Yb $4f^{14}6s^2$	71 Lu $4f^{14}5d^16s^2$

† Actinides

90 Th $6d^27s^2$	91 Pa $5f^26d^17s^2$	92 U $5f^36d^17s^2$	93 Np $5f^46d^17s^2$	94 Pu $5f^67s^2$	95 Am $5f^77s^2$	96 Cm $5f^76d^17s^2$	97 Bk $5f^97s^2$	98 Cf $5f^{10}7s^2$	99 Es $5f^{11}7s^2$	100 Fm $5f^{12}7s^2$	101 Md $5f^{13}7s^2$	102 No $5f^{14}7s^2$	103 Lw $5f^{14}6d^17s^2$

Figure 4.7 Periodic Chart Illustrating the Pattern of Electron Configurations

1. Same Energy levels

53

As mentioned earlier in this chapter, all the noble gases have stable electron configurations. It is of interest to summarize all the elements with respect to their electron configurations and study the pattern or periodicity which is established. Figure 4.7 gives a breakdown of the relationship between the electron configurations of the various elements. Notice that the elements in columns IA and IIA all have *s* orbital configurations whereas, all the elements in columns IIIA, IVA, VA, VIA, VIIA and the noble gases have *p* orbital configurations. There also is a more sophisticated pattern in the transition elements and the Lanthanide and Actinide series involving the *d* orbitals and *f* orbitals respectively. These patterns will be of use in the discussion of the various types of bonding which follows in the next chapter.

Glossary

Aufbau Principle—A principle in quantum theory which states that an electron upon entering the energy levels of an atom will occupy the lowest energy orbital at which a vacancy exists.

Excited State—An energy state which is created when an electron is in an energy level higher than predicted by the Aufbau Principle.

Ground State—The lowest possible energy state of an atom.

Hund's Rule—The principle which dictates the order of filling of orbitals of equal energy. There is a maximum number of half-filled orbitals within a given subshell.

Orbital—A region of space in an atom. A three-dimensional probability distribution for locating a given electron about the nucleus of an atom.

Pauli Exclusion Principle—The principle in quantum theory which states that no two electrons in the same atom may have the same set of quantum numbers. Relates to the spin quantum numbers.

Quantum—A discrete quantity. In energy terms, the concept that energy is released in distinct steps and not as a continual flow of energy.

Subshell—A set of like orbitals within a given energy level. The subshell is related to the "1" or secondary quantum number.

Exercises

1. List the postulates of Bohr's Atomic Theory. For each postulate state whether it is used in modern quantum mechanics or not; and if it has been altered, what is the change which has been made?

2. What is the meaning of the statement: "Energy is quantized."

3. What is the meaning of the term "orbital" as used in modern quantum mechanics?

4. Write electron configurations for the following atoms. What properties do these elements have in common?

   a. Helium   d.   Krypton

   b. Neon   e.   Xenon

   c. Argon

5. Give an example of an element that has three unpaired electrons in its ground state configuration.

6. What is the difference between "ground state" and "excited state" for a given atom?

7. What is the maximum number of electrons that each of the following may contain:

   a. Any orbital   c.   2nd Energy level

   b. 1st Energy level   d.   3rd Energy level

8. Write electron configurations for the following atoms.

   a. Sodium   d.   Chlorine

   b. Fluorine   e.   Oxygen

   c. Calcium

9. Complete the following table:

element	atomic number	electron configuration	schematic representation
a. _____	__7__	_____	
b. __B__	_____	_____	
c. _____	_____	$1s^2\, 2s^2\, 2p^6\, 3s^2$	
d. _____	_____	_____	
e. _____	_____	$1s^2\ \ 2s^2\ \ 2p^2$	

10. Complete the following table:

element	atomic number	atomic mass	number protons	number neutrons	electron configuration
a. P					
b.				5	
c.			13		
d.		4.00			
e.	16				

11. Define the following terms:

    a. Pauli Exclusion Principle—

    b. Aufbau Principle—

    c. Hund's Rule—

12. In terms of atomic structure and quantum theory, study Figure 4.7 and explain why there are eighteen elements in the fourth period of the modern periodic chart.

CHAPTER 5

The Chemical Bond

5.1 Introduction to Chemical Bonding

By definition, a chemical bond is the chemical combination of two atoms to give rise to a new substance. There are two basic points which must always be considered in discussing chemical bonding. First, *chemical bonding* is the transfer or sharing of electrons between the two bonding atoms. Second, the noble gas electron configuration is achieved in most chemical bonds. That is, the bonding atoms each achieve a completed energy level. Because of this second principle, the noble gas family is the key to chemical bonding.

For a chemical bond to occur, both bonding atoms must be in need of some adjustment in terms of electron configuration so as to achieve a noble gas configuration. This means an atom either loses, gains, or shares a given number of electrons to achieve a noble gas configuration. Whether an atom loses or gains electrons is dependent on the electron structure of the atom as well as the structure of the other chemical species to which the bond is to occur.

5.2 Covalent Bonding

From the previous section, it is apparent that atoms can combine in a variety of ways to form molecules. One type of bonding is called *covalent bonding*. Covalent bonding is the mutual sharing of electrons between the two atoms to form a stable molecule. In covalent bonding, each bonding atom has an electron configuration which is generally within four electrons of a stable bonding configuration. Since each bonding atom has a shortage of electrons, it is convenient for the two atoms to combine together in such a manner to share electrons in an attempt to achieve a noble gas configuration. Remember, in this type of bonding there is a common need for both bonding atoms to gain electrons.

Covalent bonding is characteristic of nonmetals as well as many transition metals. The bonding between two nonmetals will always give rise to a covalent bond.

5.3 Covalent Bonding—The Hydrogen Molecule

The simplest covalent bond we can discuss is that of the hydrogen molecule. The hydrogen molecule, H_2, is a diatomic molecule formed by the equal sharing of an electron between two hydrogen atoms. This is shown in Figure 5.1 using a graphic representation for the atomic and molecular orbitals involved.

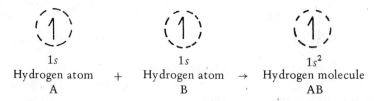

$1s$		$1s$		$1s^2$
Hydrogen atom	+	Hydrogen atom	→	Hydrogen molecule
A		B		AB

Figure 5.1. Graphic Diagram of an H_2 Molecule.

The electron configuration for a hydrogen atom is $1s^1$. If you consider the noble gas configuration, each hydrogen atom must gain one electron in order to obtain the more stable noble gas configuration. This is accomplished by having a hydrogen atom share an electron with another hydrogen atom. The end result is a hydrogen molecule with each of the atoms approaching the noble gas configuration of helium, $1s^2$.

Now consider the bonding of another diatomic molecule, S_2. The sulfur atom has the electron configuration $1s^2\,2s^2\,2p^6\,3s^2\,3p^4$. Considering the *valence electrons* or the electrons in the outer energy level, a graphic representation is illustrated in Figure 5.2.

In the case of sulfur, there is a sharing of two electrons by each atom in order to approximate a noble gas configuration or completed energy level. This is accomplished when the electrons in the $3p_y$ orbital of each atom share electrons and likewise in the $3p_z$. This is also a covalent bond and is called a double bond because there are two pairs of electrons being shared.

Sulfur atom A

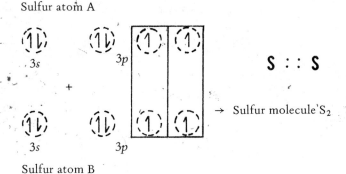

Sulfur atom B

Figure 5.2. Graphic Diagram for Sulfur Molecule, S_2.

Nitrogen atom A

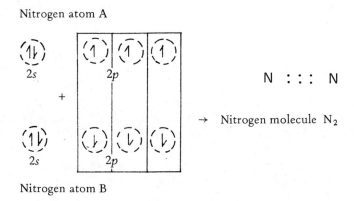

Nitrogen atom B

Figure 5.3. Graphic Diagram for the Nitrogen Molecule, N_2.

As a final example of covalent bonding, consider the nitrogen molecule, N_2. This molecule forms a triple bond as shown in Figure 5.3. A triple bond means the sharing of three pairs of electrons.

It should be evident from these examples that covalent bonding occurs when both bonding species are short valence electrons. Notice the atoms combine to form the noble gas configuration of the closest noble gas. Remember only the valence electrons are involved in the formation of the bond and that electrons must be donated from each atom not just from one atom.

Now let us consider the bonding in a molecule such as water. With the formula H_2O it is evident that one molecule of water contains two hydrogen atoms and one oxygen atom chemically bonded together. Let us see how this is accomplished.

Each hydrogen atom requires one electron to complete its energy level and the oxygen is short two electrons to complete its energy level. Therefore, the oxygen accepts both hydrogen electrons locating one in each of the half-filled p orbitals. This is illustrated in Figure 5.4.

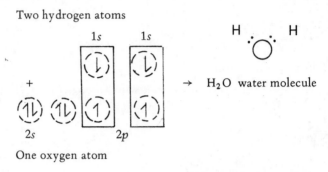

Figure 5.4. Bonding in the Water Molecule.

This is not considered a double bond but rather two single bonds because each bond between atoms is a sharing of a pair of electrons. A combination of double or triple bonds is also possible. An example of a molecule forming two double bonds would be sulfur dioxide, SO_2, and an example of a single and a triple bond in the same molecule is illustrated in HCN, hydrogen cyanide. It is important to remember that atoms may bond in a variety of ways and not just in a one-to-one correspondence as illustrated by the diatomic gases.

5.4 Ionic Bonding

When two atoms combine to form a new compound, there are many times when there is not a mutual sharing of electrons. That is, one of the bonding species has an electron configuration which exceeds the noble gas structure by up to four electrons and the other bonding species has an electron configuration which is short of the noble gas structure by up to four electrons. In terms of stability, the atom which has an excess of electrons transfers the necessary number of electrons to the atom which has a shortage of electrons so that both species approximate a noble gas configuration. Such a transfer of electrons between bonding atoms is called an *ionic bond*. An ionic bond is the transfer of electrons from one bonding atom to another to form a new chemical species.

Ionic bonding is exhibited when a metal combines with a non-metal. For example, consider the bonding of sodium and fluorine. Sodium has an excess of one valence electron and fluorine is short one electron from having a completed energy level. Thus, if one electron were transferred from the sodium atom to the fluorine atom, each atom would approximate a noble gas electron configuration. The graphic representation for this bond is illustrated in Figure 5.5.

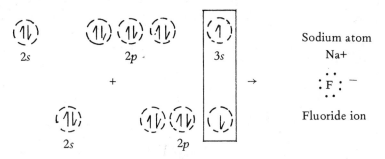

Figure 5.5. Ionic Bonding in Sodium Fluoride.

5.5 Hybridization

As mentioned in Chapter 4, the lowest energy state of an atom is not necessarily the bonding state. That is, in certain substances the bonding state will be of a higher energy than the ground state configuration. The reason for this is that an electron is promoted from one orbital to a higher orbital and this *promotion energy* is the difference in the ground state and the bonding state. An electron is promoted to create a bonding situation which is more stable and consequently, lower in energy than a bonding configuration in the ground state. The operation of forming these new bonding orbitals is called *hybridization*. The carbon atom is of primary importance in the discussion of hybridization; let us consider its behavior in some detail.

Carbon has six electrons and the ground state configuration is $1s^2\ 2s^2\ 2p^2$. However, very few stable carbon compounds are formed from this configuration. Rather, carbon compounds with four covalent bonds are much more common. This is accomplished by promoting one of the $2s$ electrons to a vacant $2p$ orbital and physically forming a new set of orbitals called hybrid orbitals. This is illustrated in Figure 5.6.

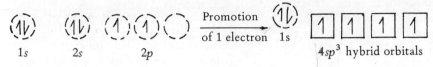

Figure 5.6. Formation of the sp^3 of the Hybrid Orbital.

It must be emphasized that the four sp^3 hybrid orbitals formed are not the normal s atomic orbital or the p atomic orbitals but rather a completely new shape as well as different physical and chemical properties. The sp^3 hybrid orbitals always give rise to a tetrahedrally shaped molecule such as methane, CH_4, with a bond angle of 109.5°. You can think of a tetrahedron as a triangular based pyramid with the carbon atom in the exact center. This is illustrated in Figure 5.7.

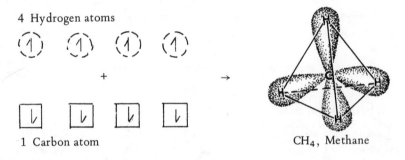

Figure 5.7. Bonding in the Methane Molecule.

Unless an exception is noted, you should assume that carbon will form the sp^3 hybrid orbitals and consider its bonding as involving four sp^3 hybrid orbitals giving rise to the tetrahedral bond angles.

There are other types of hybrid orbitals involving other atoms and giving rise to other geometric shapes but for our work, the concepts presented in the previous section will be sufficient.

5.6 Ions—Ionization Energy

In Section 5.4, the formation of ionic compounds was discussed. Let us now consider this formation in greater detail. As was pointed out, the sodium atom in sodium fluoride transferred an electron and the fluorine atom gained this electron. Because of this transfer of electrons, new chemical species called *ions* are formed. Ions are charged chemical species generally obtained by the transfer of electrons from neutral atoms. In the above case, the sodium atom transfers one electron to form a sodium ion, Na^+. The sodium ion has a positive one charge as it now contains eleven protons and ten electrons. On the other hand, fluorine gains an electron and forms a fluoride ion, F^-. The fluoride ion has a negative one charge because it contains nine protons and ten electrons.

As more and more examples of ion formation are studied, the general pattern of electron transfer will become evident. In general, the metal atoms will form metallic ions with a positive charge and the nonmetal atoms will form nonmetallic ions with a negative charge.

There are two types of ions. They are called *cations* or ions with a positive charge, and *anions* or ions with a negative charge. Table 5.1

TABLE 5.1
COMMON IONS

Cation	Name	Anion	Name
NH_4^+	Ammonium	CH_3COO^-	Acetate
H^+	Hydrogen	Br^-	Bromide
K^+	Potassium	ClO_3^-	Chlorate
Na^+	Sodium	Cl^-	Chloride
Ag^+	Silver	CN^-	Cyanide
Cu^+	Copper (I)	F^-	Fluoride
Cu^{2+}	Copper (II)	H^-	Hydride
Ca^{2+}	Calcium	OH^-	Hydroxide
Pb^{2+}	Lead (II)	I^-	Iodide
Mg^{2+}	Magnesium	NO_3^-	Nitrate
Hg^{2+}	Mercury (II)	MnO_4^-	Permanganate
Sn^{2+}	Tin (II)	CO_3^{2-}	Carbonate
Zn^{2+}	Zinc	CrO_4^{2-}	Chromate
Al^{3+}	Aluminum	O^{2-}	Oxide
Fe^{2+}	Iron (II)	SO_4^{2-}	Sulfate
Fe^{3+}	Iron (III)	S^{2-}	Sulfide
		PO_4^{3-}	Phosphate

lists some of the common cations and anions with the appropriate name and charge.

It is often of interest to know the amount of energy required to release one of the electrons from a neutral atom to form a cation. The energy required to release one of these electrons is called the *ionization potential* or *ionization energy*. The energy required to release the most loosely held electron from the atom is called the first ionization energy; the energy required to remove the second electron is called the second ionization energy and so on. Table 5.2 lists the first, second, and third ionization energies for the first twenty elements.

TABLE 5.2
IONIZATION ENERGIES IN kcal/mole.

Elements	1st	2nd	3rd
H	314		
He	567	1254	
Li	124	1744	2823
Be	215	420	3548
B	191	580	875
C	260	562	1104
N	335	683	1094
O	314	811	1267
F	402	808	1445
Ne	497	947	1500
Na	119	1091	1652
Mg	176	347	1848
Al	138	434	656
Si	188	377	772
P	254	453	696
S	239	540	807
Cl	300	549	920
Ar	363	637	943
K	100	734	1100
Ca	141	274	1181

Notice the pattern established by the values of the ionization energies. As the ionization energy increases, the stability of the ion decreases. This is especially evident if you study the values for sodium and magnesium. Notice how much larger the 2nd ionization energy is for sodium than the first ionization energy. Make the same comparison for magnesium. Notice that this time the jump in ionization energy occurs between the second and the third ionization energy. This confirms the stability of the ionic structure and the noble gas configuration for the ion. Once the noble gas structure is approximated, it takes a great deal of energy to liberate any additional electrons.

5.7 Electronegativity

As mentioned in ionic and covalent bonding, there are times when one chemical species will attract the bonding electrons more strongly than the companion species as in the case of sodium fluoride. In this case, the fluorine attracted the electrons much more than did the sodium. This property is called *electronegativity*. That is, fluorine has a greater electronegativity than sodium.

The pattern of electronegativity can be traced on a periodic chart with the highest electronegative elements found in the upper-righthand corner. By contrast, those elements with the lowest electronegativity are found in the lower-lefthand corner of the chart. In a given column, the electronegativity generally decreases as you go down the column. For a row, the electronegativity increases as you go from left to right across the chart. Table 5.3 illustrates the electronegativity for the first twenty elements.

5.8 Oxidation States

The study of the entire process of structure and bonding is to develop insight and understanding about the formation of chemical substances. One goal in this study is to be capable of writing correct chemical formulas of compounds and also be able to explain why the formula is written in a particular manner.

As a means to this end, a system for writing chemical formulas has been developed using *oxidation numbers.* The idea of oxidation numbers is an arbitrary system by which the charge on all free atoms and molecules is neutral. It is also necessary that each ion have its correct charge. Since this is an arbitrary system, there are some rules which you must follow in order to achieve the desired results.

Rule 1: The oxidation state of all free or uncombined elements is zero. Some examples would be the diatomic gases, the homoatomic elements such as S_8 or P_4, and all free metals such as Ag and Zn.

Rule 2: The oxidation state of oxygen is a negative two except in peroxides (−1).

Rule 3: The sum of the oxidation states in a molecule must be zero. In an ion, the sum of the oxidation states must equal the net charge on the ion.

Rule 4: The oxidation state for hydrogen is a positive one except in hydrides (−1).

Rule 5: The oxidation state for any monatomic ion is the same as the charge of the ion.

TABLE 5.3
ELECTRONEGATIVITY OF THE FIRST 20 ELEMENTS

Element	Electronegativity
H	2.1
He	—
Li	1.0
Be	1.5
B	2.0
C	2.5
N	3.0
O	3.5
F	4.0
Ne	—
Na	0.9
Mg	1.2
Al	1.5
Si	1.8
P	2.1
S	2.5
Cl	3.0
Ar	—
K	0.8
Ca	1.0

Rule 6: All other oxidation states are obtained mathematically using the above rules as a guide.

Fortunately, the above rules can be related to the periodic table and thus the memorization becomes very limited. Referring to the periodic table, the following principles are noted:

1. In all compounds, the elements in Group IA have an oxidation state of positive one. These elements, remember, are the alkali-metals and are therefore, very low in electronegativity and consequently form only this one oxidation state.
2. In all compounds, the elements of Group IIA have an oxidation state of positive two. Examples would be calcium and magnesium.
3. The major oxidation state for Group IIIA is a positive three although there are other minor oxidation states.
4. The members of Group VIIA have a multitude of oxidation states ranging from −1 to +7 with the most common being −1 and +5.

5. The elements of Groups IV-VIA have a variety of oxidation states and it is advisable to learn these through practice and not memorization.

Table 5.4 illustrates the writing of chemical formulas using oxidation states.

TABLE 5.4
OXIDATION STATES AND FORMULA WRITING

Molecular Formula	Symbol with Oxidation State		Name
NaCl	Na(1)	Cl(−1)	Sodium chloride
N_2	$N_2(0)$		Nitrogen gas
H_2O	H(1)	O(−2)	Water
CO_2	C(4)	O(−2)	Carbon dioxide
$CaCl_2$	Ca(2)	Cl(−1)	Calcium chloride
$FeCl_2$	Fe(2)	Cl(−1)	Iron (II) chloride
$FeCl_3$	Fe(3)	Cl(−1)	Iron (III) chloride
Cl_2	$Cl_2(0)$		Chlorine gas
H_2O_2	H(1)	O(−2)	Hydrogen peroxide
Al_2O_3	Al(3)	O(−2)	Aluminum oxide

Glossary

Anion—An ion with a negative charge.

Cation—An ion with a positive charge.

Covalent Bond—The chemical bonding of two species together with mutual sharing of the electrons.

Double Bond—When two electrons from each bonding atom combine together to form a chemical bond involving two electron pairs.

Electronegativity—A scale to measure the tendency of an element to attract electrons.

Hybridization—The process of promoting an electron or electrons from the ground energy state to a new bonding energy state. An example is the sp^3 hybrid orbitals in carbon.

Ionization Energy—The energy required to remove an electron from an atom. There is a distinct ionization energy for each electron removed.

Oxidation State—An arbitrary system of assigning numbers to chemical species so as to facilitate formula writing and equation balancing.

Promotion Energy—Energy required to move an electron from the ground state to the bonding state.

Single Bond—A chemical bond involving one pair of electrons.

Triple Bond—A chemical bond involving three electrons from each bonding atom.

Valence Electrons—Usually the electrons in the outermost, occupied energy level.

Exercises

1. Draw graphic diagrams to illustrate the bonding in the following chemical compounds.

    a. F_2

    b. LiF

    c. NaCl

    d. NO_2

    e. HCl

2. Classify each of the following as an ionic or covalently bonded compound.

    a. N_2O_4

    b. $CaBr_2$

    c. SO_2

    d. $FeCl_3$

    e. H_2

3. Describe the pattern of electronegativity as you go from left to right across row three of the periodic chart.

4. Why does magnesium have such a high 3rd ionization energy in comparison with its first two ionization energies?

5. Predict the molecular shape of the following molecules.

    a. H_2O

    b. CO_2

    c. CH_4

    d. CCl_4

    e. O_2

6. Determine the number of electrons, protons, and neutrons for the following species.

    a. Fluorine atom

    b. Fluoride ion

    c. Neon atom

    d. Sodium ion

    e. Sodium atom

7. Name, and then classify each of the following as a cation or anion.

    a. Cl^-

    b. Na^+

    c. CH_3COO^-

    d. $SO_4{}^{2-}$

    e. $NH_4{}^+$

8. Rank the following substances in decreasing order of electronegativity. That is, write the most electronegative first.

a. Br
b. Co
c. F

d. Cs
e. Li
f. S

9. Assign oxidation states to the following elements. The formulas given are correct.

a. LiF
b. $CaCl_2$
c. Hg_2Cl_2
d. $SnCl_4$
e. Br_2

f. Ag
g. $Ca(OH)_2$
h. LiH
i. SiO_2
j. $AlPO_4$

10. Write correct formulas for the following substances.

a. Iodine gas
b. Carbon disulfide
c. Acetic acid
d. Sodium carbonate
e. Neon gas

f. Ammonia
g. Ammonium ion
h. Methane
i. Aluminum oxide
j. Magnesium sulfate

CHAPTER 6

Nomenclature

6.1 Introduction

In Chapters 4 and 5 we learned how the various atoms combine to form ions and molecules. If an intelligent discussion is to follow, it is necessary that the appropriate names for the ions and molecules be learned.

There are two systems of nomenclature used in chemistry. The first, which is also the official system, is the system developed by the International Union of Pure and Applied Chemistry (IUPAC System) and is known as the *Stock System*. The second system is called the *Classical System* or derived system of nomenclature. A student must be familiar with both systems but it should be pointed out that the Stock System is the accepted system and should be used whenever possible.

In the Stock System, the oxidation state of the cations is shown in parenthesis using Roman Numerals following the name of the ion. The Classical System uses an -ous or -ic ending on the name of the cation to indicate a lower or higher oxidation state of the ion. The -ous ending is the lower oxidation state and the -ic ending is the higher oxidation state.

6.2 Nomenclature of Cations

Using the rules obtained in the previous section, the names of the various cations may be written. In general, if a cation has but one principle oxidation state, the oxidation state is not written following the name. This rule is important when naming the ion of the metals of Group I and IIA as they have but one principle oxidation state. Table 6.1 lists many of the common cations with the correct Stock System and Classical System name. You will notice that the -ous and

TABLE 6.1
NOMENCLATURE OF COMMON CATIONS

Formula	Stock Name	Classical Name
NH_4^+	Ammonium	Ammonium
Cu^+	Copper (I)	Cuprous
Cu^{2+}	Copper (II)	Cupric
H^+	Hydrogen	Hydrogen
Hg^{2+}	Mercury (II)	Mercuric
Hg_2^{2+}	Mercury (I)	Mercurous
K^+	Potassium	Potassium
Ag^+	Silver	Silver
Na^+	Sodium	Sodium
Ba^{2+}	Barium	Barium
Cd^{2+}	Cadmium	Cadmium
Ca^{2+}	Calcium	Calcium
Co^{2+}	Cobalt (II)	Cobaltous
Fe^{2+}	Iron (II)	Ferrous
Fe^{3+}	Iron (III)	Ferric
Pb^{2+}	Lead (II)	Plumbous
Mg^{2+}	Magnesium	Magnesium
Mn^{2+}	Manganese (II)	Manganous
Mn^{4+}	Manganese (IV)	Manganic
Zn^{2+}	Zinc	Zinc
Cr^{3+}	Chromium (III)	Chromic

-ic ending does not indicate a particular value. That is, one time the oxidation states might be 1 and 2 and the next time 2 and 3. This is the drawback of the Classical System. The Stock System does not have this drawback as the true oxidation state is always listed following the name.

6.3 Nomenclature of Anions

The simplest anions are those composed of but one element. In general, anions are nonmetals and if they are composed of but one element, their name will take on an -ide ending. In other words, use

TABLE 6.2
NOMENCLATURE OF SIMPLE ANIONS

Formula	Ionic Name	Formula	Ionic Name
Br^-	Bromide	N^{3-}	Nitride
C^{4-}	Carbide	O^{2-}	Oxide
Cl^-	Chloride	P^{3-}	Phosphide
F^-	Fluoride	S^{2-}	Sulfide
H^-	Hydride	$*CN^-$	Cyanide
I^-	Iodide	$*OH^-$	Hydroxide

* Exceptions to the system.

the stem name of the atom and add the -ide ending to indicate the ionic form. This procedure is illustrated in Table 6.2.

Another group of anions which must be considered are those which involve a nonmetal bonded with oxygen to form a stable anion. The procedure to remember in this case is that an -ate ending is used to name the species with the higher oxidation state and an -ite ending is used to name the lower oxidation state anion. This system is illustrated in Table 6.3.

TABLE 6.3
NOMENCLATURE OF ANIONS CONTAINING OXYGEN

Formula	Ionic Name	Formula	Ionic Name
CO_3^{2-}	Carbonate	$Cr_2O_7^{2-}$	Dichromate
SO_4^{2-}	Sulfate	NO_3^-	Nitrate
SO_3^{2-}	Sulfite	NO_2^-	Nitrite
ClO_3^-	Chlorate	AsO_4^{3-}	Arsenate
ClO_2^-	Chlorite	BrO_3^-	Bromate
ClO^-	Hypochlorite	$C_2O_4^{2-}$	Oxalate
PO_4^{3-}	Phosphate	IO_3^-	Iodate
CrO_4^{2-}	Chromate		

A group of anions which should also be treated separately are those which contain both hydrogen and oxygen along with some nonmetallic element. Also included in this group are those ions which contain more oxygen than would be predicted by the oxidation state. The best method to remember these is to study the following table as the rules can be more confusing than the actual names. Table 6.4 illustrates the formula and common name for members of this group.

TABLE 6.4
NOMENCLATURE OF SOME COMPLEX ANIONS

Formula	Name
HCO_3^-	Bicarbonate (Hydrogen carbonate)
HSO_4^-	Bisulfate (Hydrogen sulfate)
MnO_4^-	Permanganate
ClO_4^-	Perchlorate
CH_3COO^-	Acetate
O_2^{2-}	Peroxide

6.4 Nomenclature of Molecules

After becoming familiar with the naming of the various ions, the next logical step is to combine these ions together to form compounds. This can be accomplished by combining a metal ion with a nonmetal ion or the combining of one nonmetal ion with another nonmetal ion.

First, let us consider the combining of a metal ion with a nonmetallic ion to form a group of compounds called *salts*. The key to remember is that the molecular representation for salts must indicate a neutral substance. That is, the total positive charge must equal the total negative charge. This is accomplished by taking the appropriate number of ions of each species involved until the substance formed is neutral. Subscripts and parentheses are used to clarify the number of ions used.

To name the substance obtained, merely take the cation name followed by the anion name. Remember to write the positive ion first.

Example 6.1: Write the correct formula for calcium chloride.
   Solution:
   a. The symbol for the calcium ion is Ca^{2+}.
      The symbol for the chloride ion is Cl^-.
   b. The molecule must be neutral; therefore, two chloride ions are needed to neutralize one calcium ion.
   c. In equation form:

   $$Ca^{2+} + 2\,Cl^- \rightarrow CaCl_2 \text{ Calcium Chloride}$$

   d. Notice the fact that there are two chloride ions per molecule of calcium chloride. This is illustrated by using a subscript of 2 after the chloride ion.

Example 6.2: Write the correct formula for Magnesium nitrate.

Solution:
a. The symbol for the magnesium ion is Mg^{2+}.
   The symbol for the nitrate ion is NO_3^-.
b. The molecule must be neutral; therefore, two nitrate ions are needed to neutralize one magnesium ion.
c. In equation form:

$$Mg^{2+} + 2 NO_3^- \rightarrow Mg(NO_3)_2 \quad \text{Magnesium Nitrate}$$

d. Parentheses are used to clarify the fact that two nitrate ions are required to form a neutral molecule. Remember the NO_3^- acts as a single group with a negative one charge per ion and thus two nitrate ions are required to neutralize the magnesium ion.

Table 6.5 lists some of the common examples of molecules formed by the bonding of metals and nonmetals to form salts.

In the bonding of two nonmetallic atoms, the more electropositive species is written first. A prefix, indicating the number of atoms, is attached to each element when naming. Common prefixes and their numerical values are listed in Table 6.6.

TABLE 6.5
NAMES AND FORMULAS
OF SOME COMMON SALTS

Formula	Name	Formula	Name
$NaNO_3$	Sodium nitrate	$FeCl_3$	Iron (III) chloride
Na_2SO_4	Sodium sulfate	$FeSO_3$	Iron (II) sulfite
$CaCO_3$	Calcium carbonate	KNO_3	Potassium nitrate
LiH	Lithium hydride	Ag_2SO_4	Silver sulfate
$MgCl_2$	Magnesium chloride	$ZnBr_2$	Zinc bromide
$Al_2(SO_4)_3$	Aluminum sulfate	$CuClO_3$	Copper (I) chlorate

TABLE 6.6
PREFIXES AND THEIR VALUES

Prefix	Numerical Value	Prefix	Numerical Value
Mono-	1	Hexa-	6
Di-	2	Hepta-	7
Tri-	3	Octa-	8
Tetra-	4	Nona-	9
Penta-	5	Deca-	10

Table 6.7 illustrates some of the common nonmetallic molecules and their common names.

TABLE 6.7
NOMENCLATURE OF NONMETALLIC MOLECULES

Formula	Name	Formula	Name
CO	Carbon monoxide	N_2O_4	Dinitrogen tetroxide
CO_2	Carbon dioxide	PCl_3	Phosphorus trichloride
SO_2	Sulfur dioxide	PCl_5	Phosphorus pentachloride
N_2O	Dinitrogen oxide	CCl_4	Carbon tetrachloride
NO_2	Nitrogen dioxide	SO_3	Sulfur trioxide

6.5 Nomenclature of Acids

Let us now consider a group of compounds which contain hydrogen. When the hydrogen in a compound ionizes when dissolved in water, a group of compounds called *acids* is obtained. The nomenclature of acids is dependent upon the anion obtained when the acid ionizes.

Remember, when an acid ionizes, two types of ions are formed. The cation or positive ion formed is the hydrogen ion and the anion formed is the negative ion of a conjugate acid. The pattern of naming is quite systematic as can be noted in the following table. The principles are:

1. Any -ide anion will give a hydro—ic acid.
2. Any -ate anion will give an—ic acid.
3. Any -ite anion will give an—ous acid.

Table 6.8 tabulates some of the common acids and their respective names.

TABLE 6.8
NOMENCLATURE OF INORGANIC ACIDS

Formula	Name	Formula	Name
HF	Hydrofluoric acid	$HClO_3$	Chloric acid
HCl	Hydrochloric acid	$HClO_2$	Chlorous acid
HBr	Hydrobromic acid	HClO	Hypochlorous acid
HI	Hydriodic acid	HNO_3	Nitric acid
HCN	Hydrocyanic acid	HNO_2	Nitrous acid
H_2SO_4	Sulfuric acid	HIO_3	Iodic acid
H_2SO_3	Sulfurous acid	H_3BO_3	Boric acid
H_3PO_4	Phosphoric acid	$H_2C_2O_4$	Oxalic acid
H_3PO_3	Phosphorous acid	CH_3COOH	Acetic acid
$HClO_4$	Perchloric acid	H_2CO_3	Carbonic acid

6.6 Nomenclature of Bases

Another group of compounds of interest to a chemist is a group called *bases.* In the classical sense, the bases are a group of compounds formed by the associating of a metallic cation with the hydroxide ion. Thus, most bases will have a hydroxide ending. A list of some of the bases is given in Table 6.9.

TABLE 6.9
THE NOMENCLATURE OF BASES

Formula	Name	Formula	Name
NaOH	Sodium hydroxide	$Fe(OH)_2$	Iron (II) hydroxide
KOH	Potassium hydroxide	$Fe(OH)_3$	Iron (III) hydroxide
$Ca(OH)_2$	Calcium hydroxide	CuOH	Copper (I) hydroxide
$Mg(OH)_2$	Magnesium hydroxide	$Cu(OH)_2$	Copper (II) hydroxide
$Al(OH)_3$	Aluminum hydroxide	$Zn(OH)_2$	Zinc hydroxide
LiOH	Lithium hydroxide	$Mn(OH)_2$	Manganese (II) hydroxide
$Ba(OH)_2$	Barium hydroxide	$Pb(OH)_2$	Lead (II) hydroxide

The previous sections have given the basis for common chemical nomenclature. The principles presented will allow you to name most chemical compounds. However, remember there are some exceptions to any set of rules and you must be ready to adjust and remember these exceptions through practice and experience.

Glossary

Acid—In a classical sense, a compound which contains the hydrogen ion. It is better defined as a substance which donates a proton or is able to accept a species containing an electron pair.

Base—In a classical sense, a compound which contains the hydroxide ion. It is better defined as a substance which accepts a proton or is able to donate a species containing an electron pair.

Classical System—A system of nomenclature or naming of compounds which uses the -ous and -ic endings to indicate the various oxidation states.

Nomenclature—The systematic process of naming chemical substances.

Salt—A compound formed by the chemical combinations of metallic and non-metallic chemical species.

Stock System—A system of nomenclature adopted by the International Union of Pure and Applied Chemistry. This system uses the Roman Numeral to indicate the oxidation state. It is now the recommended system of nomenclature.

Exercises

1. Classify each of the following compounds as an acid, base, or salt.

    a. KOH
    b. $CaCl_2$
    c. HIO_3

    d. $PbCl_2$
    e. $Mg(NO_3)_2$

2. Name the following ions using the Stock System and the Classical System of nomenclature.

    a. Cu^+
    b. Cu^{2+}
    c. Fe^{2+}
    d. Fe^{3+}
    e. Pb^{2+}

    f. Pb^{4+}
    g. K^+
    h. Hg^{2+}
    i. Hg_2^{2+}
    j. Co^{2+}

3. Write correct formulas for the following compounds.

    a. Chlorate ion
    b. Acetate ion
    c. Sulfite ion
    d. Fluoride ion
    e. Iodate ion

    f. Borate ion
    g. Carbonate ion
    h. Oxalate ion
    i. Cyanide ion
    j. Hydride ion

4. What is the main advantage of the I.U.P.A.C. System over the Classical System of nomenclature?

5. Write the correct formula for the following ions.

    a. Ferrous ion
    b. Plumbous ion
    c. Cupric ion
    d. Mercuric ion
    e. Mercurous ion

    f. Cobaltous ion
    g. Nickel (II) ion
    h. Cuprous ion
    i. Manganese (II) ion
    j. Manganese (IV) ion

6. Name the following acids using the I.U.P.A.C. System of nomenclature.

    a. H_3PO_4
    b. H_2CO_3
    c. H_2SO_4

    d. CH_3COOH
    e. HCl
    f. HNO_3

g. $H_2C_2O_4$ i. $HClO_3$

h. HBr   j. HIO_3

7. Write formulas for the following acids.

a. Nitrous acid   d. Hypochlorous acid

b. Perchloric acid   e. Hydriodic acid

c. Hydrofluoric acid

8. Write formulas for the following bases.

a. Sodium hydroxide   d. Ferric hydroxide

b. Calcium hydroxide   e. Aluminum hydroxide

c. Magnesium hydroxide

9. Name the following salts.

a. KClO   f. NH_4I

b. Mg_3N_2 g. $Ca(CH_3COO)_2$

c. $BiCl_3$ h. CuC_2O_4

d. $CaSO_3$ i. $Al_2(CO_3)_3$

e. $KHSO_4$ j. $MnSO_3$

10. Write formulas for the following compounds.

a. Calcium oxide   f. Nickel (II) bromate

b. Magnesium perchlorate   g. Arsenic acid

c. Dinitrogen tetroxide   h. Lithium hydride

d. Silicic acid   i. Cadmium chromate

e. Sodium sulfate   j. Potassium dichromate

CHAPTER 7

The Chemical Equation

7.1 Introduction

Whenever chemists discuss a chemical change, they invariably have an abbreviated method of expressing the change. Such a method is called a *chemical equation*. A chemical equation can be written either in words or with formulas. For example, consider the chemical change when hydrogen gas and chlorine gas react to form hydrogen chloride gas. The word equation for the chemical reaction is: hydrogen gas + chlorine gas yields hydrogen chloride gas. Using chemical formulas, the chemical reaction is written:

$$H_2(g) + Cl_2(g) \rightarrow 2HCl(g)$$

Let us now consider the various forms a chemical equation can have and at the same time be aware of the information available from a balanced equation.

7.2 The Chemical Equation

As mentioned in the introduction, a word equation expresses the chemical reaction in words. A word equation is acceptable as long as no calculation or detailed information is required. Usually, however, additional information is desired and thus an equation using formulas is more desirable. Once the equation is written in formula form, information with respect to composition or structure becomes readily available. Thus, when any detailed study of a chemical reaction is required, the equation is written in terms of formulas and symbols.

7.3 Format and Procedure

When writing a chemical equation, the formulas for the various chemical species are obtained by following the rules of nomenclature. Once the formulas are obtained, it then becomes necessary to organize them together in a meaningful relationship. An equation is composed of chemical species called reactants and products. The *reactants* are the species which undergo the chemical change. The *products* are the species produced by the chemical change. By convention, the reactants are written on the left side of the equation, and the products on the right side. Various symbols are used to indicate additional information. Some of the symbols, along with their meanings, are summarized in Table 7.1.

TABLE 7.1
SYMBOLS USED IN EQUATIONS

Symbol	Meanings
(s)	solid phase of matter
(l)	liquid phase of matter
(g)	gas phase of matter
Δ	heat
(aq)	a substance dissolved in water (aqueous)
$\rightarrow$	yields
$\rightleftharpoons$	indicates a reversible reaction
=	equals

7.4 Balancing Chemical Equations

The concept behind the balancing of chemical equations is the Law of Conservation of Mass. That is, the same number of atoms of each element must be found on each side of the equation. To achieve the conservation of mass, the coefficients before the chemical species are adjusted until the same number of atoms of each species is found on each side of the equation. *Under no conditions are the formulas of the species changed to balance an equation.*

A chemical equation can be interpreted in terms of atoms and molecules or in terms of gram-atoms and moles. Consider the reaction of carbon burning in oxygen to form carbon dioxide. The equation is $C(s) + O_2(g) \rightarrow CO_2(g)$. If we were to interpret this equation, we could say that one atom of carbon reacts with one molecule of oxygen to form one molecule of carbon dioxide. Or we could say one gram-atom of carbon reacts with one mole of oxygen to form one mole of carbon dioxide.

To varify that the Law of Conservation of Mass has been obeyed, substitute masses in grams for the species. We would then say that 12 grams of carbon reacts with 32 grams of oxygen to form 44 grams of carbon dioxide. Thus, the total mass of the reactants equals the total mass of the product and the conservation of mass is verified.

The following examples illustrate the stepwise procedure for balancing equations. As one becomes experienced with the balancing of equations, certain patterns involving the coefficients will become evident. Study these patterns to become proficient at balancing equations.

Example 7.1: Balance the following equation:

$$Ca + O_2 \rightarrow CaO$$

Solution: The unbalanced equation has two oxygen atoms on the left and but one on the right. Balance the oxygen atoms by placing a coefficient of 2 before the CaO. This balances the oxygen but the calcium is no longer balanced. The calcium is balanced by putting a coefficient of two before the calcium on the left side.

$$2\,Ca + O_2 \rightarrow 2\,CaO \text{ (balanced)}$$

Example 7.2: Balance the following equation:

$$Al + O_2 \rightarrow Al_2O_3$$

Solution: The unbalanced equation has two oxygen atoms on the left and three oxygen atoms on the right. Also, there is one aluminum atom on the left and two aluminum atoms on the right side. The oxygen is the key to balancing this equation. Since both 2 and 3 are prime numbers, the way to balance the oxygen with whole numbers is to place a 3 before the O_2 on the left side and a 2 before the Al_2O_3 on the right side. The equation then becomes: $Al + 3O_2 \rightarrow 2Al_2O_3$. Notice that the oxygen is balanced with six oxygen on each side. To complete the balancing, balance the aluminum by putting a 4 before the aluminum on the left side. $4Al + 3O_2 \rightarrow 2Al_2O_3$.

Note: The 2:3 ratio as seen with the oxygen atoms is quite common in balancing equations. It can be balanced by finding the least common multiple which is 6. Therefore, there must be at least 6 oxygen atoms on each side of the equation when balanced.

Example 7.3: Balance the following equation:

$$Cu + AgNO_3 \rightarrow Ag + Cu(NO_3)_2$$

Solution: This equation involves the nitrate ion, $NO_3{}^-$. Notice that the nitrate ion does not change in structure throughout the reaction. When this occurs, it is easier to treat the entire ion as a single species rather than to split the ion into individual atoms. There are two nitrate ions on the right and but one on the left. To balance the nitrate ion, put a coefficient of 2 before the $AgNO_3$. The equation becomes: $Cu + 2AgNO_3 \rightarrow Ag + Cu(NO_3)_2$. To complete the balancing of the equation, the silver atoms must be balanced by placing a 2 before the Ag on the right side. The balanced equation is:

$$Cu + 2AgNO_3 \rightarrow 2Ag + Cu(NO_3)_2$$

Example 7.4: Balance the following equation:

$$Ca(OH)_2 + H_3PO_4 \rightarrow Ca_3(PO_4)_2 + H_2O$$

Solution: It is again possible to balance this equation by the use of ions. There are two $PO_4{}^{3-}$ ions on the right and but one on the left side. Balance the phosphate ions by placing a coefficient of 2 before the H_3PO_4.

$$Ca(OH)_2 + 2H_3PO_4 \rightarrow Ca_3(PO_4)_2 + H_2O$$

The calcium is also unbalanced with three calcium atoms on the right and one on the left side. Balance by putting a 3 before the calcium on the left side.

$$3Ca(OH)_2 + 2H_3PO_4 \rightarrow Ca_3(PO_4)_2 + H_2O$$

Finally, balance the hydrogen and oxygen. There is a total of twelve hydrogen on the left side and but two on the right side. Balance by putting a 6 before the H_2O which in turn will balance the oxygen.

$$3Ca(OH)_2 + 2H_3PO_4 \rightarrow Ca_3(PO_4)_2 + 6H_2O \text{ (balanced equation)}$$

To balance an equation, the following rules will be of aid.

1. Once the formulas for the reactants and products are correctly written, do not change them to balance the equation.
2. By convention, reactants are written on the left side and products on the right side.
3. To have an equation balanced, the total number of atoms of each element must be the same on each side of the equation.

4. All atoms and ions are balanced by inserting the proper coefficient before the molecule, ion, or atom. Remember, a coefficient placed before a substance multiplies each atom in the substance by that coefficient.
5. Make sure in balancing one element that the balance of another element is not disturbed.
6. Ions which do not change in the reaction may be balanced as a single quantity.
7. As a general rule, balance the hydrogen and the oxygen last.
8. The final coefficients of the balanced equation must be in the lowest whole number terms possible.

7.5 Types of Chemical Reactions

There are four basic types of chemical reactions that we are concerned with at this time. Table 7.2 summarizes the type of reaction along with a general equation as well as a specific example.

TABLE 7.2
TYPES OF CHEMICAL REACTIONS

Type of Reaction	General Equation	Example
Decomposition	$AB \rightarrow A + B$	$2HgO \rightarrow 2Hg + O_2$
Synthesis	$A + B \rightarrow AB$	$2Mg + O_2 \rightarrow 2MgO$
Substitution	$A + BC \rightarrow B + AC$	$Zn + H_2SO_4 \rightarrow ZnSO_4 + H_2$
Metathesis	$AB + CD \rightarrow AD + CB$	$NaOH + HNO_3 \rightarrow NaNO_3 + H_2O$

Example 7.5: Classify the following chemical reactions and the equation.

   a. $KClO_3 \rightarrow KCl + O_2$

Solution: This is a *decomposition* reaction with the single species $KClO_3$ dissociating to form KCl and O_2. To balance, notice the 3:2 ratio of the oxygen atoms. A least common multiple of 6 is necessary and is achieved by putting a 2 before the $KClO_3$ and a 3 before the O_2. The KCl is balanced with a 2 also.

   $2KClO_3 \rightarrow 2KCl + 3O_2$ (balanced equation)

   b. $H_2 + Cl_2 \rightarrow HCl$

Solution: This is a *synthesis* reaction with the two reactants combining to form a single product. To balance the equation, put a coefficient of 2 before the HCl molecule.

$$H_2 + Cl_2 \rightarrow 2HCl \text{ (balanced equation)}$$

c. $Mg + H_2SO_4 \rightarrow MgSO_4 + H_2$

Solution: This is a *substitution* reaction with the magnesium metal replacing the hydrogen in the sulfuric acid. The equation is balanced as written.

d. $NaOH + H_2CO_3 \rightarrow Na_2CO_3 + H_2O$

Solution: This is a *metathesis* reaction with the reactants changing ionic partners to form the products. This type of reaction is sometimes referred to as a double replacement reaction. To balance the equation, put a coefficient of 2 before the NaOH.

$$2NaOH + H_2CO_3 \rightarrow Na_2CO_3 + H_2O \text{ (balanced equation)}$$

7.6 Heat Effects in Reactions

Often, when discussing chemical reactions, the energy factor must be considered. That is, is heat liberated during the reaction or is heat required for the reaction to occur? For example, when the catalyst and the resin of a fiber glass cement are mixed, there is a tremendous release of heat. When heat is liberated during a chemical reaction, the reaction is said to be *exothermic*. When heat or energy is required for a reaction to occur, the reaction is said to be *endothermic*. If a reaction is exothermic, the heat or energy term is written as a product. In an endothermic reaction, the energy term is written as a reactant. For example, consider the reaction of H_2 and Br_2 to form HBr. This is an exothermic reaction as indicated by the energy term written as a product. You could also say that the reverse reaction is endothermic as energy is required for the reaction to occur. The reaction is: $H_2 + Br_2 = 2HBr + 24$ Kcal.

Glossary

Endothermic—Relating to a chemical reaction in which energy is required for the reaction to occur. Energy is absorbed during the reaction.

Equation—In terms of chemistry, a shorthand notation to illustrate a chemical reaction. An equation may be written in either words or chemical symbols.

Exothermic—A chemical reaction which liberates energy. The energy term is written as a product in the chemical equation.

Products—Those chemical species produced by a chemical change or reaction.

Reactants—Those chemical species which participate in a chemical change or reaction.

Exercises

1. Write the following word equations as balanced formula equations:

   a. Calcium carbonate → calcium oxide + oxygen

   b. Hydrogen + oxygen → water

   c. Zinc + hydrochloric acid → hydrogen + zinc chloride

   d. Acetic acid + ammonium hydroxide → ammonium acetate + water

   e. Hydrogen + bromine → hydrobromic acid

   f. Mercury (II) oxide → mercury + oxygen

   g. Aluminum oxide → aluminum + oxygen

   h. Phosphorus + chlorine → phosphorus trichloride

2. Classify each of the following equations according to Table 7.2. Balance the equations.

   a. $H_2 + I_2 \rightarrow HI$

   b. $CH_4 + O_2 \rightarrow CO_2 + H_2O$

   c. $HNO_3 + KOH \rightarrow KNO_3 + H_2O$

   d. $Pb + O_2 \rightarrow PbO$

   e. $BaI_2 + NH_4NO_3 \rightarrow Ba(NO_3)_2 + NH_4I$

3. Complete and balance the following reactions:

   a. $H_2 + O_2 \rightarrow$ _____(synthesis)

   b. $Ca\,CO_3 \rightarrow$ _____$+ CO_2$ (analysis)

   c. $Al + HCl \rightarrow$_____ + _____
      (substitution)

   d. $Ca(OH)_2 + HCl \rightarrow$_____ +_____
      (metathesis)

4. Indicate a common, everyday occurrence that is an example of an exothermic reaction. Do the same for an endothermic reaction.

5. Balance the following equations:

   a. $N_2 + H_2 = NH_3$

   b. $CO + NO_2 = CO_2 + NO$

   c. $PbI_2\,(s) = Pb^{2+}(aq) + I^-(aq)$

d. $C_2H_6 + O_2 = H_2O + CO_2$

e. $SO_2 + O_2 = SO_3$

f. $HCl + O_2 = H_2O + Cl_2$

g. $C_{12}H_{22}O_{11} + O_2 = CO_2 + H_2O$

h. $HBr + Ca(OH)_2 = CaBr_2 + H_2O$

i. $NO + O_2 = NO_2$

j. $Rb + Br_2 = RbBr$

k. $Na + H_2O = NaOH + H_2$

l. $Cl_2 + H_2O = HOCl + HCl$

m. $KNO_3 = KNO_2 + O_2$

n. $Mg(NO_3)_2 + H_2SO_4 = HNO_3 + Mg(HSO_4)_2$

o. $CH_3COOH + Al(OH)_3 = Al(CH_3COO)_3 + H_2O$

CHAPTER 8

Stoichiometry—I

8.1 Introduction to Chemical Calculations

In Chapter 2, the definitions of atoms and molecule were given and some simple chemical calculations were illustrated. Time was spent in learning the symbols for the various elements and the method of combining the various atoms together to form ions and molecules. The writing of equations has also been under study in the previous chapters.

It is now time to combine these thoughts with some mathematical concepts and observe some of the information which may be obtained through calculations based on formulas and balanced chemical equations.

8.2 Review of Gram-Atomic Weight and Gram-Molecular Weight

By definition, the *gram-atomic weight* is the weight of Avogadro's number of atoms of an element, measured in grams.

The *gram-molecular weight* is the weight of Avogadro's number of molecules of a substance. The molecular weight of a substance is the sum of the various atomic weights of the atoms which go to make-up the substance. (See Section 2.4). Remember that Avogadro's number is 6.02×10^{23} (often referred to as merely 6×10^{23}). Therefore, when expressing a gram-atomic weight or gram-molecular weight, the number of atoms or molecules involved is 6×10^{23}.

Consider the following examples of gram-atomic weight and gram-molecular weight calculations.

Example 8.1: The symbol for sulfur is S. What is the atomic weight of sulfur?

    Solution: By referring to the periodic chart, it may be noted that sulfur has an atomic weight of 32.064 amu (atomic mass units). Remember that the atomic weight on the chart is the average mass of all the isotopes of that element. Thus, no one isotope has the exact mass of 32.064 amu. However, it is the average mass and is thus used. In order to express the mass of sulfur in grams, 6×10^{23} atoms of sulfur must be considered. The mass or weight of 6×10^{23} atoms of sulfur be 32.064 grams.

Example 8.2: Calculate the gram-molecular weight of Na_2CO_3.

    Solution: Just as in Example 1, you must consider 6×10^{23} molecules of sodium carbonate if the molecular weight is expressed in grams. The gram-molecular weight is the sum of the various gram-atomic weights.

$$2\ Na\ atoms = 2 \times 23 = 46$$
$$1\ C\ atom = 1 \times 12 = 12$$
$$\underline{3\ O\ atoms = 3 \times 16 = 48}$$
$$gram\text{-}molecular\ weight = 106\ grams$$

Notice in the above example, the atomic weights have been rounded to the nearest tenth of a gram. This is often done for convenience of calculation and is acceptable for general work.

Example 8.3: Calculate the gram-molecular weight of $Ca_3(PO_4)_2$.

    Solution:

$$3\ Ca\ atoms = 3 \times 40 = 120$$
$$2\ P\ atoms = 2 \times 31 = 62$$
$$\underline{8\ O\ atoms = 8 \times 16 = 128}$$
$$gram\text{-}molecular\ weight = 310\ grams$$

8.3 The Mole and Gram-Atom

From the previous section, it has been found that Avogadro's number of particles is often used in a calculation. For example, we calculated that 6×10^{23} molecules of Na_2CO_3 weighed 106 grams and 6×10^{23} atoms of sulfur weighed 32 grams. Because of the common usage of this value, two terms are introduced to cover their meanings with respect to molecules and atoms.

A *mole* is an abbreviation or term used for "gram-molecular weight." By definition, 1 mole of a substance is 6.02×10^{23} molecules of that substance. From the previous example, one mole of Na_2CO_3 would contain 6×10^{23} molecules of Na_2CO_3 and weigh 106 grams. To calculate the number of moles of a substance present, divide the number of grams by the gram-molecular weight:

$$\# \text{moles} = \frac{\# \text{grams}}{\text{GMW}}$$

GMW = gram-molecular weight

Example 8.4: Calculate the number of moles present in 54 grams of H_2O.

Solution: First determine the gram-molecular weight and then substitute into the equation to find the number of moles.

$$\# \text{moles} = \frac{\# \text{grams}}{\text{GMW}}$$

2 H atoms = 2
1 O atom = 16
GMW = 18 g/mole

$$\# \text{moles} = \frac{54 \text{ grams}}{18 \text{ g/mole}}$$

$$\# \text{moles} = 3.0 \text{ moles}$$

A *gram-atom* is the term used for "gram-atomic weight." By definition, 1 gram-atom of a substance is 6.02×10^{23} atoms of that substance. From Section 8.2, 1 gram-atom of sulfur would contain 6.02×10^{23} atoms of sulfur and weigh 32 grams. The number of gram-atoms of a substance is calculated in much the same manner as the number of moles. That is, the number of gram-atoms is equal to the number of grams divided by the gram-atomic weight of the material.

$$\# \text{gram-atoms} = \frac{\# \text{grams}}{\text{GAW}} \qquad \text{GAW} = \text{the mass of a gram-atomic weight}$$

Example 8.5: Calculate the number of gram-atoms in 115 grams of sodium, symbol Na.

Solution: Obtain the atomic weight of sodium from the periodic chart and substitute into the equation to find gram-atoms.

$$\# \text{ gram-atoms} = \frac{\# \text{ grams}}{\text{GAW}} = \frac{115 \text{ grams}}{23 \text{ g/g-A}} = 5 \text{ gram-atoms}$$

Example 8.6: Complete the blanks in the following table.

number moles CO_2	number molecules CO_2	number C atoms	number O atoms	number gram-atoms C	number gram-atoms O
a. 1					
b. 5					
c.	1				

Solution Line (a)

Step 1: 1 mole CO_2 equals 6×10^{23} molecules of CO_2 by definition of a mole.

Step 2: The formula CO_2 indicates 1 carbon atom/molecule of CO_2.
   Therefore, $1(6 \times 10^{23}) = 6 \times 10^{23}$ C atoms.

Step 3: The formula CO_2 indicates 2 oxygen atoms/molecules of CO_2.
   Therefore, $2(6 \times 10^{23}) = 1.2 \times 10^{24}$ O atoms.

Step 4: 1 gram-atom of carbon contains 6×10^{23} C atoms.
   Therefore, 1 gram-atom of C and 2 gram-atoms of O.

Solution Line (b): Since line (a) contains 1 mole of CO_2 and line (b) contains 5 moles of CO_2, each value in line (b) must be 5 times as large as the corresponding value in line (a).

Step 1: 5 moles CO_2 = $5 (6 \times 10^{23}) = 3 \times 10^{24}$ molecules CO_2.

Step 2: 3×10^{24} molecules $CO^2 = 3 \times 10^{24}$ C atoms and 6×10^{24} O atoms.

Step 3: 1 gram-atom = 6×10^{23} atoms.
   Therefore, $\dfrac{3 \times 10^{24} \text{ C atoms}}{6 \times 10^{23} \text{ atoms/gram-atom}} = 5$ gram-atoms C,

   and $\dfrac{6 \times 10^{24} \text{ O atoms}}{6 \times 10^{23} \text{ atoms/gram-atom}} = 10$ gram-atoms O,

Solution Line (c)

Step 1: 1 molecule of CO_2 is $\dfrac{1}{6 \times 10^{23}}$ moles of CO_2.

That is, $\dfrac{1 \text{ mole}}{x} = \dfrac{6 \times 10^{23} \text{ molecules}}{1 \text{ molecule}}$.

$$x = 1.7 \times 10^{-24} \text{ moles } CO_2.$$

Step 2: In 1 molecule of CO_2 there are
  1 carbon atom and 2 oxygen atoms.

Step 3: 1 gram-atom of carbon = 6×10^{23} atoms.

Therefore, $\dfrac{1 \text{ gram-atom}}{x} = \dfrac{6 \times 10^{23} \text{ atoms}}{1 \text{ atom}}$.

$$x = 1.7 \times 10^{-24} \text{ gram atoms C.}$$

Step 4: The number of gram-atoms of
  oxygen = $2\,(1.7 \times 10^{-24}) = 3.4 \times 10^{-24}$ gram-atoms of oxygen.

8.4 Calculation of Percentage Composition

In Chapter 2, simple calculations involving the percentage composition were presented. In this chapter we shall review those early concepts and extend the information into more sophisticated areas of calculation.

The percentage composition of a compound is the weight percent of each element in the given compound. All the percentages should have a sum equal to one hundred percent. Thus, the molecular weight may be considered as 100% of the weight or total mass of the given substance. If the formula of a substance is known, the percentage composition may be calculated without need of further information.

Example 8.7: Calculate the percentage composition for H_2O.

Step 1: Determine the gram-molecular weight of H_2O.

$$\begin{array}{ll} 2 \text{ gram-atoms H} = & 2.0 \text{ g} \\ \underline{1 \text{ gram-atom O} = 16 \text{ g}} \\ \text{GMW} \qquad = 18 \text{ g/mole} \end{array}$$

Step 2: Determine the percentage of each element.

$$\% \text{ H} = \frac{\text{wt. of H}}{\text{total wt.}}(100) = \frac{2.0 \text{ g}}{18 \text{ g}}(100) = 11\% \text{ hydrogen}$$

$$\% \text{ O} = \frac{\text{wt. of O}}{\text{total wt.}}(100) = \frac{16 \text{ g}}{18 \text{ g}}(100) = 89\% \text{ oxygen}$$

In a compound which contains but two elements, the second percentage may be found by merely subtracting the first percentage from 100%. That is, 100% − 11% hydrogen = 89% oxygen. There can be slight variations in the percentage due to the rounding of numbers, but this should not exceed 0.5%. **Caution:** This method can be risky! If an error is made, there is no way of checking it.

Example 8.8: Calculate the percentage composition of $Ca(OH)_2$.

Step 1: Determine the gram-molecular weight.

$$
\begin{aligned}
1 \text{ gram-atom Ca} &= 40 \text{ g} \\
2 \text{ gram-atoms H} &= 2.0 \text{ g} \\
2 \text{ gram-atoms O} &= 32 \text{ g} \\
\hline
\text{GMW} &= 74 \text{ g/mole}
\end{aligned}
$$

Step 2: Percentage of each element:

$$\% \text{ Ca} = \frac{\text{wt. of Ca}}{\text{total wt.}}(100) = \frac{40 \text{ g}}{74 \text{ g}}(100) = 54\% \text{ calcium}$$

$$\% \text{ H} = \frac{\text{wt. of H}}{\text{total wt.}}(100) = \frac{2.0 \text{ g}}{74 \text{ g}}(100) = 2.7\% \text{ hydrogen}$$

$$\% \text{ O} = \frac{\text{wt. of O}}{\text{total wt.}}(100) = \frac{32 \text{ g}}{74 \text{ g}}(100) = 43\% \text{ oxygen}$$

There are times when the formula of the compound is not known. This creates a slightly different type of calculation. If the weights of the combining species are known, the percentage composition can be calculated by remembering that chemical reactions must obey the Law of Definite Composition.

Example 8.9: In a chemical reaction, 1.2 grams of carbon (C) were found to just react with 0.4 grams of hydrogen (H). Calculate the percentage composition.

Step 1: Calculate the mass of the substance formed.

$$
\begin{aligned}
\text{carbon} &= 1.2 \text{ grams} \\
\text{hydrogen} &= 0.4 \text{ grams} \\
\hline
\text{mass of product} &= 1.6 \text{ grams}
\end{aligned}
$$

Step 2: Determine the percentage of each element.

$$\% C = \frac{\text{wt. of C}}{\text{total wt.}} (100) = \frac{1.2 \text{ g}}{1.6 \text{ g}} (100) = 75\% \text{ carbon}$$

$$\% H = \frac{\text{wt. of H}}{\text{total wt.}} (100) = \frac{0.4 \text{ g}}{1.6 \text{ g}} (100) = 25\% \text{ hydrogen}$$

8.5 Determination of Empirical and Molecular Formulas

By definition, the *empirical formula* expresses the atoms of a compound in the smallest whole-numbered ratio possible. The empirical formula is related to the percentage composition as it is the percentage composition which dictates the ratios between the various elements. Substances may have the same percentage composition and thus the same empirical formula, yet have completely different chemical and physical properties.

A detailed study of a given substance requires that the *molecular formula,* or the formula which shows the actual number of atoms of each species in a given molecule, be known. The molecular formula can be calculated from the empirical formula by knowing the molecular weight of the substance. There are times when the empirical formula and the molecular formula are the same. If the molecular formula and the empirical formula are not the same, the molecular formula will always be an integral multiple of the empirical formula. For example, the empirical formula of a hydrocarbon might be CH and the molecular formula actually $C_6 H_6$. Notice that the molecular formula is six times larger than the empirical formula.

Example 8.10: A compound was found to contain 27.3% carbon and 72.7% oxygen. Calculate the empirical formula.

Step 1: Assume 100 grams of the compound; therefore:

$$C = 27.3\% = (100 \text{ g}) (0.273) = 27.3 \text{ g of C}$$
$$O = 72.7\% = (100 \text{ g}) (0.727) = 72.7 \text{ g of O}$$

$$\text{Total} = 100 \text{ grams}$$

Step 2: Determine the number of gram-atoms:

$$\text{gram-atoms of C} = \frac{27.3 \text{ g}}{12.0 \text{ g/g-a}} = 2.27 \text{ gram-atoms of C}$$

$$\text{gram-atoms of O} = \frac{72.7 \text{ g}}{16.0 \text{ g/g-a}} = 4.54 \text{ gram-atoms of O}$$

Step 3: Write gram-atoms in a whole number ratio by dividing each number by the smallest number of gram-atoms.

$$C_{\frac{2.27}{2.27}}O_{\frac{4.54}{2.27}} = CO_2 \text{ empirical formula}$$

Example 8.11: A hydrocarbon was found to contain 83% carbon and 17% hydrogen. Calculate the empirical formula.

Step 1: Assume 100 grams of the compound:

$$C = 83\% = (100 \text{ g}) (0.83) = 83 \text{ g of C}$$
$$H = 17\% = (100 \text{ g}) (0.17) = 17 \text{ g of H}$$

Step 2: Determine the number of gram-atoms:

$$\text{gram-atoms C} = \frac{83 \text{ g}}{12 \text{ g/g-a}} = 6.9 \text{ gram-atoms of C}$$

$$\text{gram-atoms H} = \frac{17 \text{ g}}{1.0 \text{ g/g-a}} = 17 \text{ gram-atoms of H}$$

Step 3: Determine the whole number ratio:

$$C_{\frac{6.9}{6.9}}H_{\frac{17}{6.9}} = CH_{2.47} = CH_{2.5} = C_2H_5 \text{ empirical formula}$$

In this case, the ratio upon first dividing came out to 1:2.47. This is due to rounding of the various measurements. The laws of chemical composition would indicate a combination of small whole numbers so the 2.47 is rounded to 2.5. Since 2.5 is not a whole number, the ratio is doubled to obtain the whole number ratio 2:5.

Example 8.12: If the molecular weight of the substance in example 8.11 is 58 grams/mole, calculate the molecular formula.

Step 1: Determine the formula weight of the empirical formula:

$$C_2H_5 = 29 \text{ grams/mole}$$

Step 2: Divide the molecular weight by the formula weight of the empirical formula:

$$\frac{\text{molecular weight}}{\text{empirical formula weight}} = \frac{58 \text{ g/mole}}{29 \text{ g/mole}} = 2 \text{ (formula number)}$$

The value 2 indicates the molecular formula contains twice as many gram-atoms of the elements carbon and hydrogen as did the empirical formula.

Step 3: Empirical Formula × 2 = Molecular Formula

$$C_2H_5 \times 2 = C_4H_{10} \text{ molecular formula}$$

Example 8.13: 1.4 grams of nitrogen (N) will combine with 3.2 grams of oxygen (O) to form a compound. The molecular weight of the compound is known to be 92 grams/mole. Calculate the empirical and molecular formulas.

Step 1: Calculate the mass of compound formed.

$$N = 1.4 \text{ grams}$$
$$O = 3.2 \text{ grams}$$
$$\overline{\text{product total} = 4.6 \text{ grams}}$$

Step 2: Determine the number of gram-atoms:

$$\text{gram-atoms N} = \frac{1.4 \text{ grams}}{14 \text{ g/g-a}} = 0.10 \text{ gram-atom of N.}$$

$$\text{gram-atoms O} = \frac{3.2 \text{ grams}}{16 \text{ g/g-a}} = 0.20 \text{ gram-atom of O.}$$

Step 3: Determine the empirical formula:

$$N_{\frac{0.10}{0.10}}O_{\frac{0.20}{0.10}} = NO_2 \text{ empirical formula}$$

Step 4: Determine the formula weight of the empirical formula:

$$NO_2 = 46 \text{ g/mole}$$

Step 5: Determine the ratio between the molecular and empirical formula:

$$\frac{\text{GMW}}{\text{formula weight of empirical formula}} = \frac{92}{46} = 2 \text{ (formula number)}$$

Step 6: Determine the molecular formula:

$$NO^2 \times 2 = N_2O_4 \text{ molecular formula}$$

8.6 Weight Calculations from Chemical Equations

In order to do any calculations based on a chemical equation, the equation must be balanced. Recall that an equation which is balanced shows the proper relationship for moles and gram-atoms or molecules and atoms in a chemical reaction. Consider the equation:

$$2Al + 3Cl_2 = 2AlCl_3$$

This equation could be interpreted two ways:

1. 2 atoms of aluminum plus 3 molecules of chlorine form 2 molecules of aluminum chloride.
2. 2 gram-atoms of aluminum plus 3 moles of chlorine form 2 moles of aluminum chloride.

In the first case, you are referring to atoms and molecules whereas in the second case, you are referring to gram-atoms and moles. A balanced chemical equation can be discussed at either level with meaning as long as proper usage of terms such as mole, molecule, atom, and gram-atom is made.

Equations are not balanced in terms of grams. Whenever working a weight calculation, the units should be converted to moles or gram-atoms in order to relate to the balanced equation. Problem solving based on a balanced chemical equation is often referred to as *stoichiometry.*

Example 8.14:

  a. How many moles of oxygen gas (O_2) are needed to react with sufficient H_2 to produce 3 moles of H_2O *(g)*?

Solution: Write a balanced chemical equation:

$$2H_2 + O_2 = 2H_2O$$

The molar ratio between O_2 and H_2O is 1:2. For each mole of O_2 which reacts, two moles of H_2O would be produced. Therefore, to produce 3 moles of water, 1.5 moles of O_2 would be needed. This can be shown in a proportion.

$$\frac{1 \text{ mole } O_2}{x \text{ mole } O_2} = \frac{2 \text{ moles } H_2O}{3 \text{ moles } H_2O}$$

$$2x = 3$$

$$x = 1.5 \text{ moles } O_2$$

  b. How many grams of O_2 would be required in part (a)?

Solution:

$$\text{moles} = \frac{\# \text{ grams}}{\text{GMW}}$$

$$\# \text{ grams} = (\text{moles})(\text{GMW})$$

$$\# \text{ grams} = (1.5 \text{ moles})(32 \text{ g/mole})$$

$$\# \text{ grams} = 48 \text{ grams of } O_2$$

Example 8.15: Given the following reaction, how many moles of $H_2(g)$ would be required to react with 10 moles of $Br_2(g)$?

$$H_2(g) + Br_2(g) = HBr(g)$$

Solution: Balance the equation:

$$H_2(g) + Br_2(g) = 2\ HBr(g)$$

The molar ratio between H_2 and Br_2 is a 1:1 ratio. Therefore, 10 moles H_2 would be required to react with 10 moles of Br_2. Answer: 10 moles $H_2(g)$.

b. Determine the number of grams of HBr produced if 10 moles of Br_2 reacted in part (a).

Solution: The molar ratio between Br_2 and HBr is a 1:2 ratio. Therefore, 10 moles of Br_2 would produce 20 moles HBr.

$$\text{moles} = \frac{\#\text{grams}}{\text{GMW}}$$

$\#\text{grams} = (\text{moles})\,(\text{GMW})$

$\#\text{grams} = (20\text{ moles})\,(81\text{ g/mole}) = 1{,}620\text{ grams of HBr}$

Example 8.16: In the reaction $Al + Cl_2 \rightarrow AlCl_3$, determine the number of grams of aluminum required to produce 13.35 grams of $AlCl_3$.

Step 1: Balance the equation.

$$2Al + 3Cl_2 = 2AlCl_3$$

Step 2: Determine the number of moles of $AlCl_3$ present:

$$\text{moles} = \frac{\#\text{grams}}{\text{GMW}} = \frac{13.35\text{ grams}}{133.5\text{ g/mole}} = 0.1\text{ mole AlCl}_3$$

Step 3: Determine the number of gram-atoms of aluminum required to produce 0.1 mole of $AlCl_3$. From the balanced equation, the ratio between Al and $AlCl_3$ is 1:1. Therefore, 0.1 mole of $AlCl_3$ would require 0.1 gram-atom of Al.

Step 4: Determine the mass of Al required:

$$\text{Gram-atoms} = \frac{\#\text{grams}}{\text{GAW}}$$

$\#\text{grams} = (\text{gram-atoms})\,(\text{GAW}) = (0.1\text{ gram-atoms})\,(27\text{ g/gram-atoms})$

$\#\text{grams} = 2.7\text{ grams of Al.}$

Glossary

Empirical Formula—A formula which expresses the atoms of a compound in the smallest, whole-number ratio.

Gram-Atom—A shorthand expression for "gram-atomic weight." A gram-atom of a substance is 6.02×10^{23} atoms of that substance.

Gram-Atomic Weight (GAW)—The mass in grams of Avogadro's number of atoms. This mass is based on the carbon-12 isotope.

Gram-Molecular Weight (GMW)—The mass in grams in Avogadro's number of molecules of a substance based on the carbon-12 isotope.

Mole—The number of amu in one gram; hence, the number of carbon-12 atoms in exactly 12 grams of carbon. A mole of any substance is 6.02×10^{23} molecules.

Molecular Formula—A formula which shows the actual number of atoms of each species in a given molecule. The molecular formula is always an integral multiple of the empirical formula.

Stoichiometry—A term used to describe problem-solving based on a balanced chemical equation.

Exercises

1. Determine the mass of the following:

   a. 1 gram-atom of Mg.   c. 3×10^{23} atoms of Cr.

   b. 6×10^{23} atoms of Zn.   d. 1 atom of carbon.

2. Determine the molecular mass of the following:

   a. $NaCl$ d. $CuSO_4 \cdot 5H_2O$

   b. H_3PO_4 e. CH_3COOH

   c. $Al_2(SO_4)_3$

3. Determine the number of gram-atoms in the following:

   a. 10.8 grams of Ag.   c. 6.4 grams of S.

   b. 160 grams of Be.   d. 52 grams of Li.

4. Determine the number of gram-atoms in the following:

   a. 6×10^{23} atoms of Re.   c. 24 atoms of Mn.

   b. 1.2×10^{24} atoms of Cu.   d. 1 atom of boron.

5. Calculate the number of atoms and gram-atoms in 65.37 amu of Zn.

6. Complete the following table:

number moles H_2O	number molecules H_2O	number O atoms	number gram-atoms H
_____	1.2×10^{23}	_____	_____
7	_____	_____	_____
_____	_____	_____	2×10^3

7. Determine the number of molecules in each of the following:
   a.   2 moles of Cl_2 c.   0.12 moles of HNO_3
   b.   4.4 grams of CO_2 d.   148 grams of $Ca(OH)_2$

8. Complete the following table:

number moles NO_2	number molecules NO_2	number grams NO_2
0.1	_____	_____
_____	3×10^{21}	_____
_____	_____	46

9. Complete the following table:

number gram-atoms Ni	number atoms Ni	number grams Ni
_____	_____	5.87
_____	100	_____
2.5	_____	_____

10. Calculate the percentage composition for the following:
    a.   CO   c. $FeSO_4$
    b.   AgCl   d.   LiH

11. Rank the following in order from the greatest percentage of oxygen in the compound to the lowest.
    a. $FeSO_4$ c.   AgO
    b. MnO_2 d. $KClO_3$

12. A 9.6 gram sample of carbon was heated in oxygen and converted to an oxide which weighed 14.4 grams. Calculate the percentage composition of the oxide formed.

13. A solution contains 0.1 mole NaCl, 1×10^{-2} moles of $CuNO_3$, and 0.1 mole of $NaNO_3$. Calculate the following quantities:

    a. # moles of Na^+ ion.

    b. # moles of Cl^- ion.

    c. # moles of NO_3^- ion.

    d. # of NO_3^- ions.

14. Calculate the empirical formula for each of the following compounds:

    a. 70% Fe and 30% O.

    b. 75% C and 25% H.

    c. 26.6% K, 35.4% Cr, and 38.0% O.

    d. 52.8% Sn, 12.4% Fe, 16.0% C, and 18.8% N.

15. If a stoppered flask contains 160 grams of O_2, calculate:

    a. # moles of O_2 in flask.

    b. # molecules of O_2 in the flask.

    c. # atoms of O in the flask.

16. Calculate the percentage of water in the hydrated compound $CaSO_4 \cdot 5 H_2O$.

17. 6.95 grams of a hydrated copper sulfate ($CuSO_4 \cdot xH_2O$), was heated to drive off all the water. When cool and dry, the anhydrous copper sulfate weighed 4.45 grams. Calculate the formula of the hydrate.

18. 1.60 grams of Cu was found to just react with 0.4 grams of oxygen. Determine the empirical formula.

19. If the gram molecular mass of the oxide formed in problem 18 is 79.5 grams/mole, determine the molecular formula.

20. Calculate the empirical and molecular formula of a compound which is 92.4% C and 7.6% H by weight. The gram molecular mass of the compound formed is 78 g/mole.

21. How many moles of H_2O would be produced in the following reaction if 5 moles of $Ca(OH)_2$ reacts with sufficient HCl?

$$HCl + Ca(OH)_2 \rightarrow H_2O + CaCl_2 \text{ (Unbalanced)}$$

22. According to the following equation, how many grams of HCl are required to produce 3.4 grams of H_2S?

$$FeS + HCl \rightarrow H_2S + FeCl_2 \text{ (Unbalanced)}$$

23. Consider the following equation:

$$Fe + H_2O \rightarrow Fe_3O_4 + H_2 \text{ (Unbalanced)}$$

a. How many moles of H_2O would be required to react with 1.5 gram-atoms of Fe?

b. How many grams of H_2 would be produced if 3 gram-atoms of Fe were to react with sufficient H_2O?

24. Consider the reaction:

$$Na + Cl_2 \rightarrow NaCl \text{ (Unbalanced)}$$

a. If 2.3×10^4 grams of Na react, how many moles of NaCl are produced?

b. How many grams of NaCl are produced?

25. Sulfuric acid has a density of 1.84 g/ml. It is known to be 95% H_2SO_4 by weight. (a) How many grams of sulfuric acid would be required to produce 1.8 kg. H_2O in the following reaction?

$$H_2SO_4 + NaOH \rightarrow Na_2SO_4 + H_2O \text{ (Unbalanced)}$$

b. What volume of sulfuric acid is required to produce 1.8 kg. H_2O?

CHAPTER 9

Behavior of Gases

9.1 Density of Gases

In our early discussion of matter, it was pointed out that the gas phase was undoubtedly the most difficult on which to do measurements. To facilitate experimental measurement of gases, the concept of density was employed. Density of a substance is defined as the mass of the substance divided by the volume which it occupies.

Solid materials which have a density greater than 1.00 g/ml will sink when put in water and solid substances with a density less than 1.00 g/ml will float in water. In general, gases have a density less than 1.00 g/ml and consequently, will pass up and through water. This principle is used in the experimental operation of collecting gases over water.

Since one mole of any gas will occupy the same volume at standard conditions, the density of a gas is directly proportional to the molecular weight of the gas. For example, carbon dioxide has a greater density than hydrogen because the molecular weight of carbon dioxide is 44 g/mole and that of hydrogen is 2 g/mole.

Example 9.1: Rank the following gases in decreasing order of density. That is, list the densest gas first and so on. Helium, neon, fluorine, argon, chlorine, krypton.

Solution: Since 1 mole of any gas at standard conditions occupies the same volume, the molecular weight of the gas is directly proportional to the density. That is, the greater the molecular weight, the greater the density.

	Density
He = 4 g/mole	Kr
Ne = 20 g/mole	Cl_2
F_2 = 38 g/mole	Ar
Ar = 40 g/mole	F_2 Decreasing
Cl_2 = 71 g/mole	Ne
Kr = 84 g/mole	He

Example 9.2: Calculate the density of a gas if 950 ml has a mass of 0.475 grams.

Solution:

$$\text{density} = \frac{\text{mass}}{\text{volume}}$$

$$= \frac{0.475 \text{ grams}}{950 \text{ ml}}$$

$$= 0.00050 \text{ g/ml} = 0.50 \text{ g/liter}$$

Example 9.3: The density of air is 1.293 g/liter. What is the mass of 500 ml of air?

Solution:

$$\text{density} = \frac{\text{mass}}{\text{volume}}$$

$$\text{mass} = (\text{density})(\text{volume})$$

$$= (1.293 \text{ g/l})(0.5 \text{ l})$$

$$= 0.65 \text{ grams}$$

Example 9.4: Given a density of 1.25 g/liter, calculate the volume occupied by 650 grams of carbon monoxide.

Solution:

$$\text{density} = \frac{\text{mass}}{\text{volume}}$$

$$\text{volume} = \frac{\text{mass}}{\text{volume}}$$

$$= \frac{650 \text{ grams}}{1.25 \text{ g/liter}} = 520 \text{ liters}$$

9.2 Boyle's Law

When gases are in motion and collide with the sides of the container in which they are being held, a pressure is created. The greater the number of collisions, the greater the pressure. Robert Boyle studied the relationship between pressure and volume of gases at a constant temperature. The result of his work is known as *Boyle's Law*. Boyle's Law states that the volume of a gas is inversely proportional to the applied pressure at a constant temperature. In equation form, Boyle's Law is: PV = constant (with temperature and mass constant).

When two quantities are inversely proportional, it means that when one quantity increases, the other quantity decreases accordingly. This can be illustrated with the following graph.

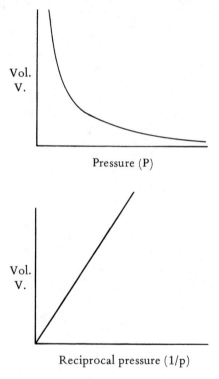

Figure 9.1. Boyle's Law—Relationship of Pressure to Volume.

Example 9.5: A sample of gas has a volume of 10 liters at 200 torr. Calculate the volume occupied by this sample if the pressure were increased to 1,000 torr.

Solution: Boyle's Law states that PV = constant. Therefore, the pressure-volume product in the initial state must equal the pressure-volume product in the final state. That is: $P_1 V_1 = P_2 V_2$

$$(200 \text{ torr}) \ (10 \text{ liters}) = (1000 \text{ torr}) \ (V_2)$$

$$\frac{(200 \text{ torr}) \ (10 \text{ liters})}{(100 \text{ torr})} = V_2$$

$$2 \text{ liters} = V_2$$

Another method of working the problem would be to notice that the pressure has increased by a factor of 5 and thus the volume must decrease by a factor of 5; therefore, the answer of 2 liters.

9.3 Charles' Law

About 100 years after Boyle discovered the relationship between pressure and volume, Jacques Charles investigated the relationship between temperature and volume at a constant pressure. The result of this work is known as *Charles' Law* which states, "The volume of a gas is directly proportional to the absolute temperature at a constant pressure." In equation form, $\frac{V}{T}$ = constant (at const. pressure and mass).

If a graph were made of this relationship, it would be similar to the one found in Figure 9.2.

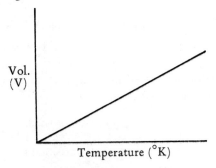

Figure 9.2. Charles' Law—Temperature vs. Volume.

Any relationship that illustrates a direct proportion would have a similar graph.

Example 9.6: A gas has a volume of 5 liters at 300°K. Assuming constant pressure, what volume is achieved if the temperature of the gas is increased to 900°K?

Solution: Using Charles' Law, $\frac{V}{T}$ = constant for both the initial and final states of the system we have:

$$\frac{V_1}{T_1} = \frac{V_2}{T_2}$$

$$\frac{5 \text{ liters}}{300°K} = \frac{V_2}{900°K}$$

$$V_2 = \frac{(5 \text{ liters}) (900°K)}{(300°K)}$$

$$V_2 = 15 \text{ liters}$$

9.4 Standard Conditions

When working with gases, it is necessary to have reference points for temperature and pressure from which all standard gas calculations can be made.

By definition, *standard temperature* is 0°C or 273°K and *standard pressure* is 1 atmosphere or 760 mm of Hg(torr). At standard conditions, one mole of any ideal gas will occupy a volume of 22.4 liters or 22,400 ml. The value, 22.4 liters, is known as the *gram-molecular volume* or molar volume of a gas at standard conditions.

9.5 The Equation of State for the Ideal Gas

If the information obtained from both Boyle's and Charles' laws are combined together, a relationship is obtained which combines the concepts of pressure, volume, and temperature in one equation. This can be shown by the following derivation.

$$\text{Boyle's Law:} \quad PV = \text{constant}$$

$$\text{Charles' Law:} \quad \frac{V}{T} = \text{constant}$$

$$\text{Therefore:} \quad \frac{PV}{T} = \text{constant}$$

If an ideal gas is considered at standard conditions, we have:

$$\frac{(1 \text{ atm}) (22.4 \text{ liters/mole})}{(273°K)} = \text{constant}$$

$$0.082 \text{ l-atm/mole } °K = \text{constant} = R$$

This constant, 0.082 liter-atm/mole-degree Kelvin is called the ideal gas constant or the molar gas constant. Thus, if all the information presented above is combined in one equation we have:

$$PV = nRT$$

where: P = pressure in atm
V = volume in liters
n = number of moles of gas
R = ideal gas constant
T = temperature in °K

This equation is called the *Equation of State for an Ideal Gas*. In order to use this equation, it is critical that each value has the proper units assigned to it. That is, the pressure must be in atmospheres, the volume in liters, and the temperature in degrees Kelvin so that the constant, R, may assume a value of 0.082 liter-atm/mole-degree Kelvin = constant = R.

Example 9.7: Calculate the volume occupied by 2.0 moles of a gas at 300°K and 2.0 atm of pressure.

Solution:

$$PV = nRT$$

$$V = \frac{nRT}{P}$$

$$V = \frac{(2.0 \text{ moles}) (0.082 \text{ l-atm/mole } °) (300°K)}{(2.0 \text{ atm})}$$

$$V = 24.6 \text{ liters} = 25 \text{ liters}$$

Example 9.8: Calculate the volume occupied by 1 mole of a gas at 27°C and 700 torr.

Solution: Convert the 27°C to °K: $(273°K) + (27°K) = 300°K$

Convert the 700 torr to atm: $\frac{700 \text{ torr}}{760 \text{ torr}} = 0.92$ atm.

$$PV = nRT$$

$$V = \frac{nRT}{P}$$

$$V = \frac{(1 \text{ mole}) (0.082 \text{ l-atm/mole } °) (300°K)}{(0.92 \text{ atm})}$$

$$V = 26.7 \text{ liters} = 27 \text{ liters}$$

Example 9.9: Calculate the molecular weight of a gas if 3.2 grams occupies a volume of 1.23 liters at 27°C and 760 torr.

Solution:

Step 1: Determine the number of moles present.

$$PV = nRT$$

$$n = \frac{PV}{RT}$$

$$n = \frac{(1.0 \text{ atm}) \ (1.23 \text{ liters})}{(0.082 \text{ 1-atm/mole}^\circ)(300^\circ K)}$$

$$n = 0.050 \text{ moles}$$

Step 2: Determine the molecular weight.

$$\# \text{ moles} = \frac{\# \text{ grams}}{G.M.W.}$$

$$G.M.W. = \frac{\# \text{ grams}}{\# \text{ moles}}$$

$$G.M.W. = 3.2 \text{ g}/.050 \text{ moles}$$

$$G.M.W. = 64 \text{ g/mole}$$

A general guideline for gas problems is that anytime you are asked to calculate the pressure, volume, temperature, moles, or the molecular weight at *a given set of conditions*, use the equation of state.

9.6 Combination Gas Law

There are times when it is necessary to change from one set of pressure-temperature conditions to a new set of conditions. Such a case is the adjustment of the volume of a gas to standard conditions. A formula has been developed for such operations. This formula, called the *combination gas law* simply uses the fact that the number of moles of gas will not change and thus, the pressure, volume, and temperature must be related. From the ideal gas law, we have:

$$PV = nRT \text{ OR } n = \frac{P_1 V_1}{R T_1}$$

If the number of moles, n, does not change we have a new relationship:

$$n = \frac{P_2 V_2}{R T_2}$$

If the two relationships are set equal to each other, we have the *combination gas law:*

$$\frac{P_1 V_1}{T_1} = \frac{P_2 V_2}{T_2}$$

Since R is in both expressions, it factors out of the equation.

When adjusting to standard conditions, call the 'values with a subscript of 2 your standard values and the values with a subscript of 1 your experimental values. A general rule to follow is that anytime you have to *change from one set of conditions to a new set of conditions* use the combination gas law.

Example 9.10: A gas has a volume of 10 liters at 27°C and 700 torr. Calculate the volume at Standard Conditions.

Solution:

$$\frac{P_1 V_1}{T_1} = \frac{P_2 V_2}{T_2}$$

$$\frac{(700 \text{ torr}) (10 \text{ liters})}{(300°\text{K})} = \frac{(760 \text{ torr}) \; V_2}{(273°\text{K})}$$

$$V_2 = \frac{(700 \text{ torr}) (10 \text{ liters}) (273°\text{K})}{(760 \text{ torr}) (300°\text{K})}$$

$$V_2 = 8.37 \text{ liters}$$

Note: Since this is a proportion, the units must be the same in each ratio.

Example 9.11: A gas has a volume of 11.2 liters at standard conditions. What volume will the gas occupy at 127°C and 2 atm of pressure?

Solution:

$$\frac{P_1 V_1}{T_1} = \frac{P_2 V_2}{T_2}$$

$$\frac{(1 \text{ atm}) (11.2 \text{ liters})}{(273°\text{K})} = \frac{(2 \text{ atm}) \; V_2}{(400°\text{K})}$$

$$V_2 = \frac{(1 \text{ atm}) (11.2 \text{ liters}) (400°\text{K})}{(2 \text{ atm}) (273°\text{K})}$$

$$V_2 = 8.2 \text{ liters}$$

9.7 Dalton's Law of Partial Pressures

In the early 1800's John Dalton stated his law of partial pressures for a mixture of gases. He stated, "The total pressure in a gaseous system is equal to the sum of the pressures of each of the gases in the system." In equation form:

$$P_{total} = P_1 + P_2 + P_3 + \ldots\ldots\ldots$$

This law is of prime importance when more than one gas is involved in a system. In order to use the gas law equations, we must know the actual pressure of the gas under study. This value can be determined very easily if one realizes that the total pressure in the system may be adjusted to equal the barometric pressure. The partial pressures of the gases involved must equal the barometric pressure.

Example 9.12: Determine the pressure of a gas collected over water at 27°C if the barometric reading is 758.4 mm of Hg. Assume all water levels are the same. Vapor pressure of H_2O at 27°C = 26.7 mm of Hg. (Remember: 1 torr = 1 mm of Hg.)

    Solution:

$$P_{total} = P_1 + P_2$$
$$P_1 = P_{total} - P_2$$
$$P_1 = 758.4 \text{ mm} - 26.7 \text{ mm} = 731.7 \text{ torr}$$

The vapor pressure of water can be determined by referring to the Table of Vapor Pressure as a function of temperature in the appendix.

9.8 Graham's Law of Diffusion

Also in the early 1800's, Thomas Graham studied the rate at which a gas will diffuse or move through a given amount of space. He discovered that the greater the density of a gas, the slower it diffused. He then stated this relationship as a comparison of rates between gases as:

$$\frac{R_1}{R_2} = \sqrt{\frac{d_2}{d_1}} \quad \text{where:} \quad \begin{array}{l} R = \text{rate} \\ d = \text{density} \end{array}$$

As was studied in Section 9.1, the density of a gas and the mass are directly proportional. Therefore, the rate of diffusion may also be related to the molecular mass of a gas. That is:

$$\frac{R_1}{R_2} = \sqrt{\frac{m_2}{m_1}} \quad \text{where:} \quad \begin{array}{l} R = \text{rate} \\ m = \text{mass} \end{array}$$

Notice that this law gives a comparison of rates and not specific values for a certain gas. Your answers should always be expressed in a simple ratio of the rates.

Example 9.13: Carbon dioxide and nitrogen are mixed together in a container. Which will diffuse more rapidly? Determine the ratio.

Solution: Since the molecular weight of carbon dioxide is 44 g/mole and nitrogen is 28 g/mole, the nitrogen will diffuse at a faster rate.

$$\frac{R_{N_2}}{R_{CO_2}} = \sqrt{\frac{m_{CO_2}}{m_{N_2}}}$$

$$= \sqrt{\frac{44 \text{ g/mole}}{28 \text{ g/mole}}}$$

$$= \sqrt{1.57}$$

$$= 1.25$$

The N_2 will diffuse 1.25 times faster than the CO_2.

9.9 Kinetic-Molecular Theory of Gases

From the previous study of gases, it has been evident that certain properties and conditions have been assumed. We talk of an ideal gas yet no gas behaves exactly like an ideal gas. However, the properties which we use to compare gases to ideal behavior have been summarized into a theory called the *Kinetic-Molecular Theory of Gases*. The basic postulates of this theory are:

1. Gases consist of small, independent, particles. Thus, gases are quite compressible due to the large amount of space between the particles.
2. Gas molecules neither attract nor repel each other.
3. Molecules move in straight lines. When there is a collision between molecules, it is a perfectly elastic collision with no change in the net energy.
4. The average kinetic energy of all gases under the same conditions is equal.

In addition to these four postulates, two derivations are also considered as guidelines in the discussion of gases. They are:

1. Ideal gases must obey Boyle's and Charles' laws with respect to pressure, temperature, and volume.

2. Graham's law of diffusion and Dalton's law of partial pressure must also be followed in accordance with ideal gas properties.

Many gases meet these criteria at normal temperatures and pressures. It is at extreme temperatures and pressures that deviation from ideal behavior occurs. Such properties as compressibility, and diffusibility change greatly at higher pressures as well as at extremely low temperatures.

For our work, we can assume ideal behavior for the gases involved. In this course the extreme conditions are not used.

Glossary

Barometer—An instrument used to measure the barometric pressure.

Boyle's Law—Assuming ideal behavior, the volume of a gas sample is inversely proportional to the applied pressure at a constant temperature.

Charles' Law—Assuming ideal behavior, the volume of a gas sample is directly proportional to the absolute temperature at a constant pressure.

Combination Gas Law—Assuming ideal behavior, an equation relating the pressure, volume and temperature of a gas sample. In equation form:

$$\frac{P_1 V_1}{T_1} = \frac{P_2 V_2}{T_2}$$

Dalton's Law of Partial Pressures—The total pressure of a gaseous system is equal to the sum of the partial pressures of the several gases in the system.

Equation of State for an Ideal Gas—An expression relating pressure, volume, temperature, and the number of moles of a gas. In equation form: $PV = nRT$.

Graham's Law of Diffusion—Relates the rate of diffusion of gases with respect to their density and molecular mass. In equation form:

$$\frac{r_1}{r_2} = \sqrt{\frac{d_2}{d_1}} \text{ OR } \frac{r_1}{r_2} = \sqrt{\frac{m_2}{m_1}}$$

Gram-Molecular-Volume—The volume occupied by one mole of a gas. The molar volume of a gas is 22.4 liters at standard conditions.

Standard Pressure—Equivalent to 1 atm or 760 torr or 760 mm of Hg.

Standard Temperature—Equivalent to $0°C$ or $273°K$.

Torr—Unit of pressure equivalent to 1 mm of Hg.

Exercises

1. Calculate the density of a gas if 1.25 liters has a mass of 0.55 grams.

2. Rank the following gases in order of decreasing density: fluorine, helium, nitrogen, hydrogen, carbon dioxide.

3. What volume will 250 grams of air occupy at standard conditions? Density of air = 1.29 g/liter.

4. Compare the rates of diffusion for oxygen and carbon dioxide. Which diffuses faster and what is the ratio?

5. The barometric pressure is 758.2 torr. Calculate the partial pressure of a gas collected over water at 25°C.

6. A sample of gas has a volume of 7 liters at 250 torr. Calculate the volume occupied by this gas if the pressure were increased to 600 torr at a constant temperature.

7. A gas has a volume of 3.2 liters at 350°K. Assuming constant pressure, what is the volume if the temperature is decreased to 100°K?

8. Calculate the volume occupied by 12 moles of CO_2 at standard conditions.

9. Calculate the volume occupied by 3 moles of O_2 at 300°K and 600 torr.

10. What is the resultant pressure if 5 moles of N_2 is found in a 12 liter container at 30°C?

11. Calculate the molecular mass of a gas if 5.26 grams occupies a volume of 2.63 liters at 127°C and 740 torr.

12. A gas sample has a volume of 12 liters at 5°C and 730 torr. Calculate the volume at standard conditions.

13. A gas sample has a volume of 3.2 liters at standard conditions. What volume will the gas occupy at experimental conditions of 22°C and 756 torr?

14. A gas sample has a volume of 16.5 liters at 25°C and 0.95 atm. Calculate the volume at standard conditions.

15. Assuming ideal behavior, how many moles of gas are present in problem 14?

16. If the gas in problem 14 were nitrogen, N_2, how many grams of nitrogen are present?

17. A gas sample has a volume of 11.5 liters at 20°C and 0.85 atm. Calculate the volume at standard conditions.

18. If the initial density of the gas in problem 17 is 0.75 g/liters, determine the mass of gas present.

19. A gas sample has a volume of 67.2 liters at standard conditions. How many moles of the gas are present?

20. Assume the gas in problem 19 is sulfur dioxide, SO_2; how many grams of SO_2 are present?

CHAPTER 10

The Chemistry of
Liquids and Solids

10.1 Introduction

In previous chapters we have discussed the various properties of solids, liquids, and gases. In Chapter 9, the kinetic molecular theory for gases was discussed. It should be pointed out, however, that all three states of matter can be discussed in terms of the kinetic molecular theory. General observations as to the properties of the three states of matter are usually made. That is, we can say that molecules are less closely held in gases than in liquids or solids. Or, the density of a gas is less than the density of a corresponding liquid or solid. Most comparisons made of matter usually involve a change in temperature and pressure. The various states of matter are important entities in themselves, but most chemical observations are made when there is a change of state involved.

10.2 Fusion

Fusion is the process of converting a substance from a solid state to a liquid state at the melting point of the given substance. In very simple terms, fusion is the process of melting.

When a substance freezes into a solid, a certain amount of energy is released. This release of energy is caused by the reduction in potential energy of the molecules which compose the substance. Molecules in a solid have very little movement due to the fact that the vibrational, rotational, and translational energy components are small due to the rigid configurations of the solid.

When a substance melts, or fuses, just the opposite case would be true. That is, there is an increase in the internal energy as the molecules become more active. The molecules in a liquid have greater freedom of movement and consequently, the opportunity to vibrate, translate, and rotate is greater. This results in a larger total energy value.

In order to perform calculations, it is necessary to designate the amount of energy required to convert a substance from a solid to a liquid state. The energy required to change one mole of a substance from the solid state to the liquid state at the melting point of the substance is known as the *molar heat of fusion.* The value of the molar heat of fusion is dependent on the substance under study. Some common values are listed in Table 10.1.

TABLE 10.1
MOLAR HEATS OF FUSION OF
SELECTED SUBSTANCES

Substance	Heat of Fusion (cal/g)
$CaCl_2$	54.3
CO_2	45.3
H_2	14.0
O_2	3.3
Pb	5.5
N_2	6.1
Ag	26.0
NaCl	124.0
S	13.2
H_2SO_4	24.0
Sn	13.8
H_2O	79.7
CH_3COOH	44.7
C_6H_6	30.3
C_2H_5OH	24.9
CH_4	14.5
T.N.T.	22.3
WAX (Bees')	42.3

10.3 Vaporization

Just as we discussed the phase change between solids and liquids, we must also consider the phase change between liquids and gases. *Vaporization* is the process of changing a substance from the liquid state to the gas state. Vaporization is an endothermic process. That is, it requires energy to occur. Evaporation would be another name for vaporization. The reverse operation of vaporization is condensation. *Condensation* is the process of changing from a gas to a liquid state.

The reason energy is required for vaporization is that a gas has even greater molecular activity than a liquid and consequently, the molecules have larger potential energy components. To convert a substance from the liquid state to the gas state, enough energy must be supplied to allow the molecules to absorb the necessary energy to translate, rotate, and vibrate at the higher energy rate.

Just as in the case of converting from a solid to a liquid, there is a standard for the conversion of a liquid to a gas. The *molar heat of vaporization* is the energy necessary for one mole of a liquid to be converted to a gas at the boiling point of the substance.

10.4 Specific Heat and Heat Capacity

We have now discussed the changes of state, but it is often necessary to discuss the energy relationship for the heating or cooling of a substance in a given phase. The quantity of energy (heat) required to raise the temperature of one gram of a substance one centigrade degree is defined as the *specific heat*. In general the specific heats of the elements are inversely proportional to the respective atomic weights. It is convenient to discuss this relationship in terms of moles as well. *Molar heat capacity* is the energy required to raise one mole of a substance one centigrade degree. Usually the energy is expressed in calories and thus the units for specific heat are cal/gram °C and for heat capacity cal/mole/°C.

Table 10.2 lists some of the common specific heats and molar heat capacities.

10.5 Heating and Cooling Curves

A very common experiment in chemistry is to measure the changes involved in the conversion of a solid substance to a liquid and a gas by heating at a constant pressure. The graphical representation of the data is called a heating curve for the substance. Let us consider the heating of one gram of water from −20°C to 120°C. Figure 10.1 illustrates the heating curve for this operation.

Notice that temperature is plotted as a function of energy which is expressed in calories. In portion (a) of the graph, the solid ice is being heated to 0°C or to the point of fusion. The only phase present at this time in the system is the solid phase. It should be noted that the specific heat of ice is 0.45 cal/gram/°C.

TABLE 10.2
SPECIFIC HEATS AND
MOLAR HEAT CAPACITIES (CONSTANT PRESSURE)

Substance	Specific Heat (cal/g)	Molar Heat Capacity (cal/mole)
Cu	0.092	5.9
Sn	0.054	6.4
Pb	0.031	6.3
Ni	0.105	6.2
NH_3	0.502	8.5
$CaCl_2$	0.164	18.2
Fe_2O_3	0.148	23.7
H_2O_2	0.471	16.0
MnO_2	0.152	21.6
$AgNO_3$	0.146	24.8
NaCl	0.204	11.9
H_2SO_4	0.239	23.4
CH_3COOH	0.487	29.2
C_6H_6	0.287	22.4
$C_6H_{12}O_6$	0.275	49.5
C_2H_5OH	0.232	10.7
O_2	0.218	7.0
H_2	3.389	6.8

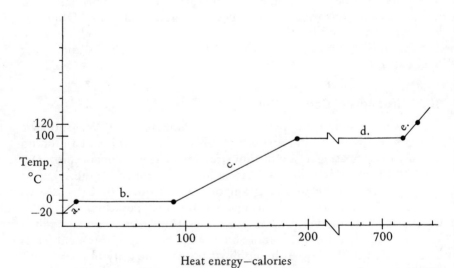

Figure 10.1. Heating Curve for Water.

Portion (b) of the curve represents the phase change between solid and liquid. The energy required to perform this operation is the heat of fusion for water which is 80 cal/gram. This means that for every gram of water changed from a solid to a liquid, 80 calories of heat is required.

Portion (c) of the curve represents the heating of the water in the liquid phase. Because the boiling point or point of vaporization is 100°C, the liquid water must be heated 100 centigrade degrees. Since the specific heat of liquid water is 1 cal/gram/°C, it takes 100 calories to change 1 gram of water from 0°C to 100°C.

Portion (d) of the curve represents the phase change between liquid and gas at the point of vaporization or the boiling point. The heat of vaporization for water is 540 cal/g/°C. Therefore, it takes 540 calories to convert 1 gram of liquid water at 100°C to vapor at the same temperature.

Portion (e) of the curve illustrates the heating of water in the gas phase. The specific heat for water vapor is 0.48/cal/g/°C. This means that for each gram of water vapor raised 1°C there is a requirement of 0.48 calories.

If the total energy for the conversion of 1 gram of water at −20°C to vapor at 120°C were calculated, the value would be 738.6 calories. This is calculated as follows:

Heating of ice from −20°C to 0°C = (.45) (20) =	9	cal
Heat of fusion =	80	cal
Heating of liquid from 0° to 100°C = (1) (100) =	100	cal
Heat of vaporization =	540	cal
Heating of gas from 100° to 120°C = (.48) (20) =	9.6	cal
Total =	738.6	cal
≈	740	cal

Notice how the use of specific heats is necessary to determine the number of calories required in heating any of the three phases. This is always the case. You must know the specific heats in order to determine the number of calories required. Consider the following examples of calorimetric calculations.

Example 10.1: Determine the number of calories necessary to convert 1 gram of water at 20°C to 110°C.

Solution:

Heat water from $20°$ to $100°C$ = (1) (80)   =   80   cal
Heat of vaporization   = 540   cal
Heat vapor from $100°$ to $110°C$ = (.48) (10) =   4.8 cal
                                    Total = 624.8 cal
                                        $\approx$ 620   cal

Example 10.2: Calculate the number of calories necessary to change 20 grams of water from $-40°C$ to $150°C$.

Solution: Determine the number of calories required for one gram and then multiply your answer by 20 grams.

For 1 gram:

Heat ice from $-40°$ to $0°C$ = (.45) (40) =   18 cal
Heat of fusion   =   80 cal
Heat liquid from $0°$ to $100°C$ = (1) (100)   = 100 cal
Heat of vaporization   = 540 cal
Heat gas from $100°$ to $150°C$ = (.48) (50) =   24 cal
                                Total = 762 cal
                                $\approx$ 760 cal

For 20 grams:

(20 grams) (762 calories/gram) = 15240 cal = 15.24 kcal
                                    $\approx$ 15   kcal

Example 10.3: Calculate the number of calories necessary to convert 2 moles of water from ice at $0°C$ to vapor at $100°C$.

Solution: 2 moles of water is 36 grams of water, so determine the number of calories required for one gram and multiply by 36 grams.

Heat of fusion   =   80 cal
Heat liquid from $0°$ to $100°C$ = 100 cal
Heat of vaporization   = 540 cal
                      Total = 720 calories/gram

For 36 grams or 2 moles:

(36 grams) (720 cal/gram) = 25920 cal = 25.92 kcal
                                $\approx$ 26   kcal

A cooling curve is the reverse operation of a heating curve. That is, it illustrates the process of a substance being changed from the gas phase to the liquid phase to the solid phase. Energy is released in this operation. Figure 10.2 illustrates the cooling curve for ammonia.

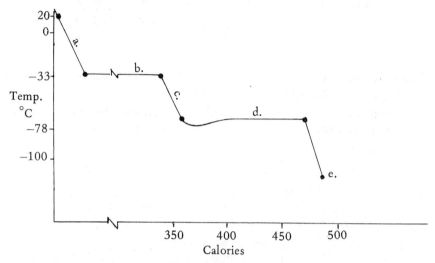

Figure 10.2. Cooling Curve for Ammonia.

Plateau (b) is the phase change between gas and liquid and plateau (d) is the phase change between liquid and solid. The dip in plateau (d) is caused by the phenomena of supercooling. When a substance fuses, the temperature may fall below the true melting point and is thus said to be supercooled. As the internal energy stabilizes, the substance returns to its normal melting point. For the temperature range of 20°C to −100°C notice that the total amount of energy released is 491 calories/gram NH_3.

A heating and cooling curve can be drawn for any substance as long as we know the temperature range and the specific heats or energy involved. These curves will always have this characteristic shape with the two plateaus illustrating the phase changes.

10.6 Phase Diagrams

As mentioned when we studied gases, temperature and pressure have a direct effect on the state of a chemical system. Under standard conditions of temperature and pressure, water will boil or vaporize at 100°C and freeze or fuse at 0°C. However, by adjusting the temperature and pressure of a system, phase changes can occur. A

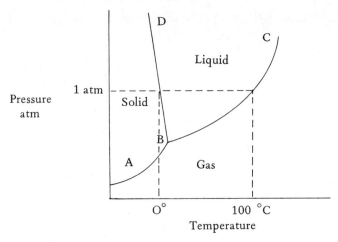

Figure 10.3. Phase Diagram for Water.

relationship between temperature, pressure, and the states of matter is called a *phase diagram.* Figure 10.3 illustrates the phase diagram for water.

Study the phase diagram in Figure 10.3. The line segment AB is the equilibrium line between solid and vapor phases at the range of temperature and pressure listed. Line segment BD is the division line between the solid and liquid phases for water and line segment BC is the division between liquid and vapor phases.

The given conditions of temperature and pressure determine what phase or phases are present. For example, at 1 atmosphere of pressure we can follow the phase changes across the graph as we increase the temperature. Notice that at 0°C we are at the solid-liquid phase change and at 100°C we are at the liquid-gas phase change.

Example 10.4: What phase or phases are present at 1.5 atmospheres of pressure and 60°C?

Solution: Locate 1.5 atm on the vertical axis and 60°C on the horizontal axis. Extend the two lines until they intersect. You should be in the *liquid phase.*

The main use of a phase diagram is that you can determine the conditions necessary for a particular phase of matter or you can study the effects of temperature and pressure on the phase of a substance.

One point of major importance in Figure 10.3 is point B. This point is called the *triple point.* All three phases of matter are present at this and only at this point. Every substance has a unique triple point.

10.7 Structure of Crystals

When the solid phase of matter is discussed in detail, the shape and characteristics of crystals becomes important. Crystalline solids can be grouped into four general classifications according to the type of particles involved in the crystal formation. Any crystal is a rigid, fixed array of particles, so it is the different ways in which the particles can be arranged that is of importance. Table 10.3 summarizes the four types of crystalline solids.

TABLE 10.3
CLASSIFICATION OF CRYSTALS

Type of Crystal	Particles Involved	Examples
Ionic	Cations and anions	$NaCl$, $CaCO_3$
Metallic	Atoms with delocalized electrons	Na, Mg
Covalent	Atoms	Diamond, graphite
Molecular	Molecules	CO_2, H_2O

Sodium chloride is an example of an ionic crystal lattice. That is, the sodium ions and the chloride ions alternate in a cubic arrangement. Sodium is an example of a metallic solid which is a systematic arrangement of sodium atoms in a body centered cubic arrangement. The electrons in a metallically bonded crystal are said to be delocalized which means that they are free to flow from atom to atom in the crystal. The two forms of carbon, graphite and diamond, have essentially a hexagonal arrangement with the diamond being extremely distorted due to pressure and temperature changes during formation. CO_2, or dry ice, is a molecular solid in which the actual molecules of CO_2 are arranged in a definite array. Figure 10.4 illustrates the structures of the various types of crystals.

When we talk of the chemical formula for crystals, we are not referring to the actual chemical bonding but rather to the empirical or simplest ratio of the particles involved. For example, we say the

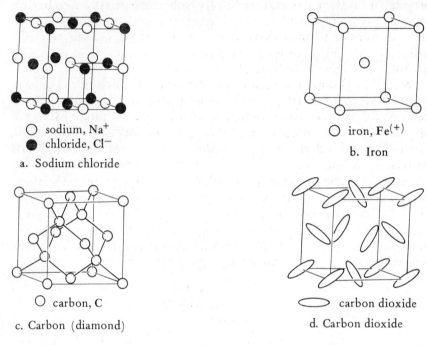

O sodium, Na$^+$
● chloride, Cl$^-$

a. Sodium chloride

b. Iron

O iron, Fe$^{(+)}$

O carbon, C

c. Carbon (diamond)

carbon dioxide

d. Carbon dioxide

Figure 10.4. Structure of Crystals.

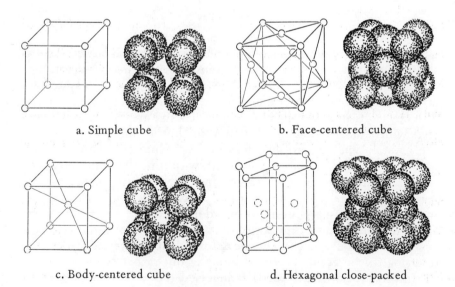

a. Simple cube

b. Face-centered cube

c. Body-centered cube

d. Hexagonal close-packed

Figure 10.5. Arrangement of Crystal Lattices.

formula of sodium chloride is NaCl which means that a number of sodium ions and chloride ions are bonded together in a one-to-one ratio. It does not mean there is one molecule of sodium chloride existing as an entity in itself. Molecular type crystals are unique in that they may exist as individual molecules.

When we discuss the various crystal arrangements, it is convenient to study the actual geometrical arrangement of the particles in the crystal. Four common arrangements or patterns which often result are called the simple cube, the face-centered cube, the body-centered cube, and the hexagonal close-packed. These patterns are illustrated in Figure 10.5.

Notice how these arrangements are incorporated into the total geometric structure. For example, study the sodium chloride structure. The chloride ions are in a face-centered cube arrangement with the sodium ions filling the holes or slots in the arrangement. All of these basic geometric arrangements in Figure 10.5 can be found in some of the common crystal arrangements.

Glossary

Calorie—A unit of energy defined as the amount of energy required to raise the temperature of one gram of water one centigrade degree.

Condensation—The chemical process of changing from the vapor state to the liquid state of matter.

Fusion—The process of changing from the solid state to the liquid state of matter.

Molar Heat Capacity—The amount of energy expressed in calories necessary to raise the temperature of one mole of a substance one centigrade degree.

Molar Heat of Fusion—The energy required to convert one mole of a substance from the solid to the liquid state.

Molar Heat of Vaporization—The amount of energy required to change one mole of a substance from the liquid state to the gaseous state.

Phase Diagram—A graphical representation of the relationship of the states of matter with temperature and pressure.

Specific Heat—The energy necessary to raise the temperature of one gram of a substance one centigrade degree.

Sublimation—The chemical process of going directly from the solid to the gas phase of matter. The liquid state is not involved.

Triple Point—The temperature at which all three phases of matter may exist simultaneously. It is defined with respect to temperature and a given pressure.

Vaporization—The process of changing from the liquid state to the gaseous state of matter.

Exercises

1. List three unique properties of solids, liquids, and gases which are common to only that particular phase of matter.

2. Ice is less dense than water. Explain this fact in terms of crystal structure and the properties of solids and liquids.

3. Why are some solids conductors of electricity and others nonconductors?

4. Explain why supercooling may occur when a pure substance freezes.

5. In the potassium chloride crystal, how many adjacent chloride ions does each potassium ion have? Assume a rock-salt lattice.

6. If all the fluoride ions were removed from sodium fluoride, what crystal lattice pattern would remain? Assume a rock-salt lattice initially.

7. Determine the amount of energy necessary to convert 1 gram of water at $5°C$ to vapor at $105°C$.

8. Calculate the number of calories necessary to change 50.0 grams of water at $-30°C$ to vapor at $200°C$.

9. In an insulated chemical system, 15 grams of water at $40°C$ was added to 25 grams of ice at $0°C$. When equilibrium is reached, what phase or phases are present and what is the resultant temperature?

10. Determine the amount of energy necessary to boil $(100°C)$ 6 cups of coffee from an initial temperature of $25°C$. Assume 1 cup of coffee contains 150 ml of coffee and has the same heat capacity as water.

11. How many grams of ice would be required to cool the 6 cups of boiling coffee in problem 10 to $80°C$? Assume the ice is at $0°C$ initially.

12. Given a pressure of 0.25 atm and a temperature of $20°C$, what phase or phases of water would be present? Refer to Figure 10.3.

13. Lead has a molar heat capacity of 6.3 cal/mole, a specific heat of 0.045 cal/gram/$°C$, and a heat of fusion of 5.5 cal/gram. The melting point of lead is $327°C$. Starting with an initial temperature of $27°C$, calculate the number of calories necessary to convert 1.0 kg of lead to a liquid at its melting point.

14. After the lead in problem 13 is melted, it is poured into a mold

made of nickel which has a specific heat of 0.105 cal/g. The nickel mold is 50 cm × 20 cm × 100 cm and weighs 2.5 kg. The initial temperature of the mold is 25°C. What is the maximum temperature achieved upon the addition of the lead?

15. Suppose the molten lead in problem 13 was poured into 5 liters of water at 20°C. What would be the resultant temperature of the water?

CHAPTER 11

Solution ·Chemistry

11.1 Terminology

Solution chemistry is the study of chemical reactions which occur in solution or in the liquid media. Since many substances are soluble, it is of major importance in beginning chemistry to be able to perform calculations involving solutions.

A solution has at least two components. One of the components is the *solute* or the substance that is present in the smaller proportion. The other component is the *solvent*. In general, the solvent is the substance present in the greater proportion. When you combine the solute and the solvent together, you have a *solution*.

There can be many different components of a solution. One of the most common is the dissolving of a solid solute in a liquid solvent. However, this is not the only method of forming a solution. Such processes as gas solute dissolved in a liquid solvent or a gas solute dissolved in a gas solvent are also solutions. By working with the three phases of matter, nine different combinations of solutions can be formed. It should be noted that some of these possibilities are rather remote but all are theoretically possible. A student must realize that a solution is more than just a solid dissolved in a liquid.

11.2 Molality

When discussing solutions, it is often necessary to know the exact relationship between the solute and the solvent. Such a relationship is shown in *molality*. Molality is defined as the number of moles of solute per kilogram of solvent. In equation form:

$$m = \text{molality} = \frac{\text{moles of solute}}{\text{kg of solvent}}$$

Thus, molality is one method of expressing concentration in a solution.

Example 11.1: Determine the molality of a solution found by adding 117 grams of NaCl to 500 grams of water.

Solution:

Step 1: Determine # of moles.

$$\text{moles NaCl} = \frac{\#g}{GMW} = \frac{117}{58.5} = 2 \text{ moles}$$

Step 2: Determine molality.

$$m = \frac{\text{moles of solute}}{\text{kg of solvent}} = \frac{2 \text{ moles}}{0.5 \text{ kg}} = 4 \text{ molal}$$

Example 11.2: Determine the molality of a solution formed by adding 18.0 grams of glucose, $C_6 H_{12} O_6$, to 500 ml $H_2 O$.

Solution:

Step 1: Determine the number of moles of solute.

$$\text{moles } C_6 H_{12} O_6 = \frac{\#g}{GMW} = \frac{18.0 \text{ g}}{180 \text{ g/mole}} = 0.100 \text{ moles}$$

Step 2: Determine the molality.

$$m = \frac{\# \text{ moles of solute}}{\text{kg of solvent}} = \frac{0.100 \text{ mole}}{0.500 \text{ kg.}} = 0.200 \text{ molal}$$

11.3 Freezing and Boiling Point Effects

Whenever an impurity is added to a substance, there is a change in the resultant freezing point and boiling point of the solution. For a nonvolatile solute, the freezing point will be lowered and the boiling point will be elevated in all cases. The actual temperature change involved depends on the amount of solute added as well as the chemical nature of both the solute and the solvent. Each substance has what is referred to as a molal freezing point depression constant, and a molal boiling point elevation constant. Table 11.1 lists some constants for some of the common chemical substances.

TABLE 11.1
FREEZING AND BOILING POINT CONSTANTS

Solvent	Boiling Point Constant B°C/Mole/Ks	Freezing Point Constant F°C/Mole/Kg
Acetic acid	3.07	3.90
Benzene	2.53	4.90
Carbon Tetrachloride	5.03	8.60
Ethanol	1.22	1.85
Ethyl ether	2.02	1.72
Phenol	3.56	7.40
Water	0.51	1.86

Since the temperature change is dependent on the molality of the substance, there is a relationship between molality and temperature change. For the freezing point the equation is:

$$\Delta T = Fm \qquad \text{where } \Delta T = \text{Temperature change}$$
$$F = \text{Freezing pt. constant}$$
$$m = \text{molality of the solution}$$

If we were to consider a boiling point change, the equation becomes:

$$\Delta T = Bm \qquad \text{where } \Delta T = \text{Temperature change}$$
$$B = \text{Boiling point constant}$$
$$m = \text{molality}$$

Example 11.3: Determine the change in the freezing point if 11.7 grams of NaCl is added to 100 ml of H_2O. The molal freezing point constant is 1.86° C/kg/mole.

Solution:

Step 1: Determine # moles of solute.

$$\text{\# moles NaCl} = \frac{\text{\#g}}{GMW} = \frac{11.7 \text{ g}}{58.5 \text{ g/mole}}$$
$$= 0.200 \text{ moles}$$

Step 2: Determine ΔT.

$$\Delta T = Fm$$

$$\Delta T = F \frac{\text{\# moles solute}}{\text{kg solvent}}$$

$$\Delta T = (1.86°\text{C/molal}) \frac{(0.200 \text{ mole})}{(0.100 \text{ kg})}$$

$$\Delta T = 3.72°\text{C}$$

11.4 Molarity

Another method of expressing concentration in solutions is to discuss concentration in terms of moles of solute and volume of solution. Such a relationship is referred to as *molarity*. Molarity is defined as the number of moles of solute per liter of solution. In equation form:

$$M = \text{molarity} = \frac{\#\,\text{moles of solute}}{\text{liter of solution}}$$

It is important to notice that the denominator of this expression is in terms of *solution* rather than *solvent* as was the case in working with molality.

Molarity is a very convenient means of expressing concentration for it uses the previously developed concept of the mole and also the total volume of solution which is easily determined by measurement. Also, molarity can be directly related to balanced chemical equations because chemical equations are also balanced in terms of moles.

Example 11.4: Determine the molarity of a solution formed by adding 0.20 moles of $CaCl_2$ (s) to enough water to make 500 ml of solution.

Solution:

$$M = \frac{\#\,\text{moles solute}}{\text{liter of solution}}$$

$$M = \frac{0.20\ \text{moles}\ CaCl_2}{0.500\ \text{liters}} = 0.40\ \text{moles/liter} = 0.40\ M$$

Notice that the volume of the solution must be expressed in *liters*. The units for molarity, therefore, are always moles/liter.

Example 11.5: Determine the molarity of a solution formed by adding 17 grams of $AgNO_3$ (s) to enough water to make 2 liters of solution.

Solution:

Step 1:

$$\#\,\text{moles}\ AgNO_3 = \frac{\#\,\text{grams}}{GMW}$$

$$\#\,\text{moles}\ AgNO_3 = \frac{17\ \text{g}}{170\ \text{g/mole}}$$

$$= 0.10\ \text{mole}$$

Step 2:

$$M = \frac{\text{\# moles of solute}}{\text{liter of solution}}$$

$$M = \frac{0.10 \text{ moles}}{2 \text{ liters}}$$

$$M = 0.05 \text{ moles/liter}$$

Example 11.6: Given a 3.0 molar solution of $Mg(NO_3)_2$, determine the number of grams of $Mg(NO_3)_2$ present in 1.5 liters of solution.

Solution: Determine the number of moles present and then calculate the number of grams.

Step 1: Determine number of moles.

$$M = \frac{\text{moles}}{\text{liter}}$$

$$\text{\# moles} = (M) \text{ (liters)}$$
$$= (3.0 \ M) \text{ (1.5 liters)}$$
$$= 4.5 \text{ moles}$$

Step 2: Determine # of grams.

$$\text{moles} = \frac{\text{\# g}}{GMW}$$

$$\text{\# g} = \text{(moles)} \ (GMW)$$
$$= (4.5 \text{ moles}) \ (148 \text{ g/moles})$$

$$\text{\# g} = 666 \text{ g} \approx 670 \text{ grams}$$

Example 11.7: State, in words, how you would prepare 2.5 liters of a $3.0M$ solution of Na_2SO_4 from solid sodium sulfate.

Solution: Determine the number of moles of Na_2SO_4 required, then determine the number of grams and express your answer in sentence form using the given volume.

Step 1: Determine the number of moles.

$$M = \frac{\#\text{ moles solute}}{\text{liter of solution}}$$

$$\#\text{ moles} = (M)\,(\text{liters})$$

$$= (3.0\ M)\,(2.5\text{ liters})$$

$$= 7.5\text{ moles}$$

Step 2: Determine the number of grams.

$$\#\text{ moles} = \frac{\#\text{ grams}}{GMW}$$

$$\#\text{ grams} = (\#\text{ moles})\,(GMW)$$

$$= (7.5\text{ moles})\,(142\text{ g/mole})$$

$$= 1065\text{ grams} \approx 1.1\text{kg}$$

Step 3: Write sentence—Weigh out 1.1 kg of Na_2SO_4 and add enough water to obtain a total volume of 2.5 liters.

11.5 Normality

A method of expressing concentration which is also based on the volume of solution is normality. *Normality* is defined as the number of gram-equivalent weights of solute per liter of solution. In equation form:

$$\text{Normality} = N = \frac{\#\text{ GEW of solute}}{\text{liter of solution}}$$

The denominator in this expression should cause no problem as it is merely the volume of solution expressed in liters. The numerator, however, can be quite confusing if not studied carefully.

If a person were to compare molarity and normality, there would be a similarity in the use of the term number of moles in molarity and the number of gram-equivalent weights in normality calculations. Thus, the number of gram-equivalent weights is equal to the number of grams present divided by the gram-equivalent weight. In equation form:

$$\#\text{ GEW} = \frac{\#\text{ of grams}}{GEW}$$

The *number of gram-equivalent* weights and the *gram-equivalent* weight do not have the same meaning. The number of gram-equivalent weights is a quantity or rather a certain number of things, whereas the gram-equivalent weight is a mass expression usually expressed in grams.

At this point, you might wonder how to obtain the gram-equivalent weight of a substance. The gram-equivalent weight is dependent on the chemical reaction in which the substance is involved. If the substance under study is an acid, the gram-equivalent weight is found by dividing the gram-molecular weight by the number of hydrogen ions liberated in the reaction. If the reaction involves a base, the gram-equivalent weight is found by dividing the gram-molecular weight by the number of hydroxide ions involved in the reaction. Finally, if the reaction involves oxidation and reduction, the gram-equivalent weight is found by dividing the gram-molecular weight of the substance by the number of electrons transferred in the reaction. This may seem complicated, but in practice it becomes quite straight-forward and easy to calculate. Table 11.2 summarizes the formulas used in normality calculations and in parenthesis you will find terms that are often used as shortcuts or abbreviations for the true term.

TABLE 11.2
SUMMARY OF NORMALITY EXPRESSIONS

$$N = \text{Normality} = \frac{\text{\# gram-equivalent weights}}{\text{liter of solution}}$$

$$\text{\# gram-equivalent weights} = \frac{\text{\# grams}}{\text{gram-equivalent weight}}$$

or

equivalent weights

or

equivalents

$$\text{gram-equivalent weight} = \frac{\text{gram-molecular weight}}{(\text{\# H}^+) \text{ or}}$$

or

equivalent weight

$(\text{\# OH}^-)$ or

or

$(\text{\# e}^-)$

equivalent mass

Example 11.8: Determine the gram-equivalent weight of the acid and the base in the following reaction:

$$2H_3PO_4 + 3Ca(OH)_2 = Ca_3(PO_4)_2 + 6\,H_2O$$

Solution:

Step 1:

$$\text{the gram-equivalent weight of } H_3PO_4 = \frac{\text{gram-molecular weight}}{\# H^+ \text{ exchanged}}$$

$$= \frac{98 \text{ g/mole}}{3 \, H^+ \text{/eq/mole}}$$

$$\text{GEW} = 32.7 \text{ g/GEW}$$

Step 2:

$$\text{the gram-equivalent weight of } Ca(OH)_2 = \frac{\text{gram-molecular weight}}{\# OH^- \text{ exchanged}}$$

$$= \frac{74 \text{ g/mole}}{2 \, OH^- \text{/eq/mole}}$$

$$\text{GEW} = 37 \text{ g/GEW}$$

Example 11.9: Determine the gram-equivalent weight of the iron metal in the following reaction:

$$Fe + CuSO_4 = FeSO_4 + Cu$$

Solution:

Step 1:

$$\text{the gram-equivalent weight of } Fe = \frac{\text{gram-molecular weight}}{\# e^- \text{ exchanged}}$$

$$= \frac{56 \text{ g/mole}}{2 \, e^- \text{/eq/mole}}$$

$$\text{GEW} = 28.0 \text{ g/GEW}$$

Note: The oxidation state of Fe on the left side of the equation is zero and on the right side of the equation is +2. Therefore, 2 electrons have been exchanged.

Example 11.10: Determine the number of gram-equivalent weights of Fe involved in Example 11.9 if it is known that 9.35 grams of Fe has reacted.

Solution:

Step 1: From Example 11.9, the gram-equivalent wt of Fe was found to be 28.0 g/GEW.

Step 2: # gram-equivalent weights $= \dfrac{\# \text{ grams}}{\text{GEW}}$

$$= \dfrac{9.35 \text{ g}}{28.0 \text{ g/GEW}}$$

$$= 0.334 \text{ GEW's}$$

Example 11.11: Determine the normality of a solution of NaOH prepared by dissolving 4.0 grams of solid NaOH in enough water to make 250 ml of solution.

Solution: $N = \dfrac{\# \text{ GEW's}}{\text{liter of solution}}$

Step 1: gram equivalent wt of NaOH

$$\text{GEW} = \dfrac{GMW}{\# \text{OH}^- \text{exchanged}}$$

$$= \dfrac{40 \text{ g/mole}}{1 \text{ OH}^-}$$

$$= 40 \text{ g/GEW's}$$

Step 2: # GEW's

$$\# \text{GEW's} = \dfrac{\# \text{g}}{\text{GEW}}$$

$$= \dfrac{4.0 \text{ g}}{40 \text{ g/GEW}}$$

$$\text{GEW's} = 0.10 \text{ GEW's}$$

Step 3: Normality

$$N = \dfrac{\# \text{GEW's}}{\text{liter}} = \dfrac{0.10 \text{ GEW}}{.25 \text{ liters}} = 0.40 \, N$$

11.6 Titration Calculations

As mentioned in Chapter 7, one type of chemical reaction involved the process of neutralization, or the reaction of an acid and a base to reach an endpoint. The physical process of such a reaction is called a *titration* and involves the use of burets. Burets are calibrated tubes which allow for very accurate measurement of volumes. The set-up for a titration is shown in Figure 11.1.

By definition, the equivalence point of a titration is the point at which the number of gram-equivalent weights of one titrated species

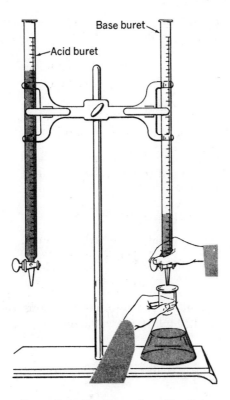

Figure 11.1. The Set-up for a Titration.

equals the number of gram-equivalent weights of the other species involved in the titration. This fact gives rise to a great number of mathematical manipulations and calculations. They are summarized in the following steps:

1. # GEW of species A = # GEW of species B
2. (Normality) (liters of solution A) = (Normality) (liters of solution B)
3. $\dfrac{\text{\# grams } A}{\text{GEW } A} = \dfrac{\text{\# grams } B}{\text{GEW } B}$

All three of these relationships are equalities and thus may be set equal to each other as the need arises in various problems. Titrations happen to be one area in which normality is much more convenient to use than molarity. The main reasons for this are the three relationships just illustrated. Let us now see how these relationships can be used in problem solving.

Example 11.12: Calculate the volume of 1.0N NaOH necessary to neutralize 25 ml of 3.0N HCl.

Solution:

$$V_{acid} N_{acid} = V_{base} N_{base}$$
$$(25 \text{ ml}) (3.0 \text{ } N) = (V) (1.0 \text{ } N)$$
$$75 \text{ ml} = V$$

Example 11.13: Determine the normality of an unknown acid if 20 ml of the acid is required to neutralize 30 ml of 4N KOH.

Solution:

$$V_{acid} N_{acid} = V_{base} N_{base}$$
$$(20 \text{ ml}) (N) = (30 \text{ ml}) (4 \text{ } N)$$
$$20N = 120$$
$$N = 6.N$$

Example 11.14: Calculate the number of grams of HIO$_3$ (s) necessary to neutralize 420 ml of 3.0M NaOH.

Solution:

$$\frac{\# \text{ grams of acid}}{\text{GEW acid}} = V_{base} N_{base}$$
$$\frac{\# \text{ grams of acid}}{176 \text{ g/GEW}} = (.421) (3.0 \text{ } N)$$
$$\# \text{ grams of acid} = (.421) (3.0 \text{ } N) (176 \text{ g/GEW})$$
$$= 221.76$$
$$\# \text{ grams of acid} = 222 \text{ grams} \approx 220 \text{ grams}$$

11.7 Dilution Calculations

As a special type of problem, the process of dilution is the simplest if considered with the proper formulas. The following examples are only for dilution processes; or in other words, when it is necessary to lower the concentration of a given substance. They do not have universal use in problem solving. Care must be taken in their use.

Example 11.15: Calculate the concentration of 200 ml of 6M HCl if it is diluted to 800 ml.

Solution:

$$(Volume_1)\,(Concentration_1) = (Volume_2)\,(Concentration_2)$$
$$(200\ ml)\,(6M) = (800\ ml)\,(C_2)$$
$$\frac{1200}{800}\,M = C_2$$
$$1.5M = C_2$$

Example 11.16: Tell how you would prepare 200 ml of 6.0 M HNO$_3$ from 16 M HNO$_3$.
Solution:

$$V_1\,C_1 = V_2\,C_2$$
$$(200\ ml)\,(6.0\ M) = V_2\,(16\ M)$$
$$\frac{1200}{16}\,M = V_2$$
$$75\ ml = V_2$$

Directions: Add 75 ml of 16M HNO$_3$ to 125 ml of H$_2$O to make 200 ml of 6.0 M HNO$_3$.

Glossary

Molality—An expression of concentration of a solution in terms of moles of solute per kilogram of solvent.

Molarity—An expression of concentration of a solution in terms of moles of solute per liter of solution.

Normality—An expression of concentration of a solution in terms of gram equivalent weights per liter of solution.

Solute—A component of a solution. It is the component which is present in smaller proportion in a solution. A solute may be a solid, liquid, or a gas.

Solution—A homogeneous mixture composed of a solute and a solvent.

Solvent—A component of a solution. It is the component which is present in greater proportion in the solution. A solvent may be a solid, a liquid, or a gas.

Titration—The physical experimental process of measuring the volumes of two reacting reagents in solution. Very often, it is the neutralization of an acid and a base.

Exercises

1. Determine the molality of a solution formed by adding 17 grams of AgNO$_3$ to 600 grams of water.

2. Determine the number of grams of solute in each of the following solutions:

   a. 200 ml of 0.1 N $AgNO_3$ d. 800 ml of 16 M HNO_3

   b. 1 liter of 0.5 M $Mg(OH)_2$ e. 500 ml of 0.5 N HCl

   c. 1.5 liters of 0.2 M NaOH

3. Determine the molality of the following solutions:

   a. 19 grams of $MgCl_2$ (s) dissolved in 100 grams of H_2O

   b. 15.4 grams of CCl_4 dissolved in 78 grams of C_2H_5OH

   c. 7.1 grams of Cl_2 (g) dissolved in 200 grams of H_2O

4. Calculate the change in the freezing point if 15.6 grams of C_2H_5OH is added to 300 ml of H_2O. The molal freezing point constant of water is 1.86° C/molal.

5. Determine the molarity of the following solutions:

   a. 2.13 grams of $Al(NO_3)_3$ in 400 ml of solution

   b. .17 grams of $AgNO_3$ in 700 ml of solution

   c. 10.6 grams of Na_2CO_3 in 1.5 liters of solution

6. Calculate the gram equivalent weight of the acid and base in each of the following reactions:

   a. $2HCl + Ca(OH)_2 \rightarrow CaCl_2 + 2H_2O$

   b. $H_3PO_4 + 2KOH \rightarrow K_2HPO_4 + 2H_2O$

   c. $H_2SO_4 + 2NaOH \rightarrow Na_2SO_4 + 2H_2O$

7. Calculate the equivalent mass of the metal in each of the following reactions:

   a. $Mg + 2HCl \rightarrow MgCl_2 + H_2$

   b. $Zn + H_2O \rightarrow ZnO + H_2$

   c. $2Al + 6HCl \rightarrow 2AlCl_3 + 3H_2$

8. State how you would prepare each of the following solutions:

   a. 200 ml of 0.05 N KOH from solid KOH

   b. 1.5 liters of 1.2 M KCl from solid KCl

   c. 400 ml of 0.75 M $CuSO_4$ from solid $CuSO_4 \cdot 5H_2O$

9. State how you would prepare each of the following solutions:

   a. 150 ml of 0.25N HCl from 2 M HCl

   b. 500 ml of 1.5 M H_3PO_4 from 6 M H_3PO_4

   c. 1.5 liters of 6 M NaOH from 12 M NaOH

10. How much water must be added to 200 ml of a solution containing 17.0 grams of $AgNO_3$ to form a 0.05 M solution?

11. Determine the number of milliliters of 1 M KOH necessary to neutralize 150 ml of 5 M H_2SO_4.

12. Calculate the normality of an unknown base if 25 ml of the base is neutralized by 40 ml of 0.50N acetic acid, CH_3COOH.

CHAPTER 12

Stoichiometry—II

12.1 Review of the Mole

In many of the previous chapters, a discussion of various mathematical calculations based on chemical equations has been made. The purpose of this chapter is to combine all the previous calculations into a single chapter so as to give a feeling of continuity to problem solving.

The basis for all chemical calculations is a balanced equation. Since chemical equations are balanced in terms of moles or molecules, it is logical to assume that the basis of all chemical calculations is also the mole.

Remember that 1 mole of any substance is equal to 6.02×10^{23} particles. In chemical terms, 1 mole of a substance is equivalent to 1 gram molecular weight of the substance which in turn is equal to 6.02×10^{23} molecules. Mathematically, the gram-atom is treated the same way as the mole. The only difference is that the gram-atom refers to atoms and the mole refers to molecules. Example 12.1 illustrates these relationships.

Example 12.1: Interpret the following balanced chemical equation in terms of moles and gram-atoms, molecules and atoms, and grams:

$$2AgNO_3 + Cu \rightarrow Cu(NO_3)_2 + 2Ag$$

a. 2 moles $AgNO_3$ + 1 gram-atom Cu = 1 mole $Cu(NO_3)_2$ + 2 gram-atoms Ag

b. 1.2×10^{24} molecules $+ 6 \times 10^{23}$ atoms $= 6 \times 10^{23}$ molecules $+ 1.2 \times 10^{24}$ atoms

c. 340 grams $+$ 64 grams $\qquad = 188$ grams $+ 216$ grams

In each of the following sections, a classic type of chemical calculation is considered. By studying the examples, a pattern for solving each type of calculation can be developed.

12.2 The Mole-Mole Calculation

When working this type of problem, you are given the number of moles of one substance and asked to calculate the number of moles of another substance based on a balanced chemical equation. It should be emphasized that *the equation must be balanced*. Also, remember that chemical equations are balanced in terms of gram-atoms or moles.

Example 12.2: How many moles of $Cu(NO_3)_2$ will be produced when 10 moles of $AgNO_3$ reacts according to the equation:

$$2AgNO_3 + Cu \rightarrow Cu(NO_3)_2 + 2Ag?$$

Step 1: Check to see that the equation is balanced. In this case, it is balanced as given as it is evident from the balanced equation that 2 moles of $AgNO_3$ will yield 1 mole of $Cu(NO_3)_2$.

Step 2: Set up a proportion to obtain your final answer. From balanced equation–2 moles $AgNO_3 \rightarrow$ 1 mole $Cu(NO_3)_2$; therefore, the ratio is 2:1. In terms of 10 moles of $AgNO_3$:

$$\frac{10 \text{ moles } AgNO_3}{2 \text{ moles } AgNO_3} = \frac{x \text{ moles } Cu(NO_3)_2}{1 \text{ mole } Cu(NO_3)_2}$$

$$2x = 10$$
$$x = 5 \text{ moles of } Cu(NO_3)_2$$

Example 12.3: Silver oxide decomposes according to the following equation:

$$Ag_2O(s) \rightarrow Ag(s) + O_2(g).$$

How many moles of O_2 would be produced if 5 moles of Ag_2O decomposes?

Step 1: Balance the equation.

$$2\,Ag_2O \rightarrow 4\,Ag + O_2$$

Step 2: Determine the molar ratio between the silver oxide and the oxygen. From the balanced equation, the ratio between Ag_2O and O_2 is 2:1.

Step 3: Set up a proportion to obtain your answer.

$$\frac{5 \text{ moles } Ag_2O}{2 \text{ moles } Ag_2O} = \frac{x \text{ moles } O_2}{1 \text{ mole } O_2}$$

$$2x = 5$$

$$x = 2.5 \text{ moles } O_2$$

Example 12.4: Calculate the number of gram-atoms of Fe required to react with 6 moles of O_2 according to the equation:

$$Fe + O_2 \rightarrow Fe_2O_3$$

Step 1: Balance the chemical equation.

$$4\,Fe + 3O_2 \rightarrow 2\,Fe_2O_3$$

Step 2: Determine the ratio between the gram-atoms of iron and moles of O_2. From the balanced equation, the ratio of $Fe:O_2$ is 4:3.

Step 3: Set up a proportion to obtain your answer.

$$\frac{x \text{ gram-atoms Fe}}{4 \text{ gram-atoms Fe}} = \frac{6 \text{ moles } O_2}{3 \text{ moles } O_2}$$

$$3x = 24$$

$$x = 8 \text{ gram-atoms of Fe}$$

The mole-mole calculation is the basis for all future chemical calculations and *must be mastered before attempting to work other types of calculations.*

12.3 The Mole-Mass Calculation

The key to solving the mole-mass calculation is to realize that you must first perform a mole-mole calculation and then determine the number of grams. In other words, when asked to determine the number of grams of a substance, you *first must determine the number of moles* of that substance produced. You then change from moles to grams with the standard formula:

$$\# \, moles = \frac{\# \, grams}{GMW}$$

Example 12.5: Determine the number of grams of NH_3 produced if 5 moles of N_2 (g) reacts with sufficient H_2 (g).

Step 1: Balance the chemical equation.

$$N_2(g) + 3 H_2(g) \rightarrow 2 NH_3(g)$$

Step 2: Determine the molar ratio between N_2 and NH_3. From the balanced equation, the desired ratio is 1:2.

Step 3: Determine the number of moles of NH_3 produced when 5 moles of N_2 reacts.

$$\frac{5 \text{ moles } N_2}{1 \text{ mole } N_2} = \frac{x \text{ moles } NH_3}{2 \text{ moles } NH_3}$$

$$x = 10 \text{ moles } NH_3 (g)$$

Step 4: Determine the number of grams of NH_3 (g) produced.

$$\# \, moles = \frac{\# \, g}{GMW}$$

$$\# \, g = (\# \, moles) \, (GMW)$$

$$\# \, g = (10 \text{ moles}) \, (17 \text{ g/mole})$$

$$\# g = 170 \text{ grams } NH_3$$

Example 12.6: Determine the number of moles of Cu_2S formed when 6.4 grams of sulfur reacts with sufficient copper.

Step 1: Write a balanced chemical equation.

$$2 Cu + S \rightarrow Cu_2 S$$

Step 2: Determine the molar ratio between S and Cu_2S. From the balanced equation, the desired ratio is a 1:1 ratio.

Step 3: Determine the number of moles of S.

$$\# \, moles = \frac{\# \, g}{GMW}$$

$$= \frac{6.4 \text{ g}}{32 \text{ g/mole}} = 0.2 \text{ moles S}$$

Step 4: Determine the number of moles Cu_2S produced from 0.2 moles S.

$$\frac{0.2 \text{ moles S}}{1 \text{ mole S}} = \frac{x \text{ moles } Cu_2 S}{1 \text{ mole } Cu_2 S}$$

$$x = 0.2 \text{ moles } Cu_2 S$$

No matter what the order of calculation is, you must eventually use the mole-mole calculation. This should be uppermost in your mind whenever carrying out any chemical calculation. Whenever you are given grams of a substance, you should immediately realize that it must be converted to moles of the substance. *Think moles or gram-atoms! "Remember the Mole."*

12.4 The Mass-Mass Calculation

The mass-mass calculation involves a balanced chemical equation in which the mass of one substance is given and the mass of another substance is the desired answer. Since equations are balanced in terms of moles rather than grams, you must change the mass expression to moles and then obtain the desired molar relationship before converting back to a mass expression.

Example 12.7: How many grams of ammonia, NH_3 will be produced if 1.4 grams of nitrogen, N_2 reacts with sufficient H_2?

Step 1: Write a balanced equation.

$$3\,H_2 + N_2 \rightarrow 2\,NH_3$$

Step 2: Determine the number of moles of N_2 present.

$$\#\,moles = \frac{\#\,grams}{GMW}$$

$$\#\,moles\ N_2 = \frac{1.4\ g}{28\ g/mole}$$

$$= 0.05\ moles\ N_2$$

Step 3: Determine the number of moles of NH_3 produced from 0.05 moles N_2.

$$\frac{0.05\ moles\ N_2}{1\ mole\ N_2} = \frac{x\ moles\ NH_3}{2\ moles\ NH_3}$$

$$x = 0.10\ moles\ NH_3$$

Step 4: Determine the number of grams of NH_3

$$\#\,moles = \frac{\#\,g}{GMW}$$

$$\#\,g = (\#\,moles)\,(GMW)$$

$$\#\,g = (0.10\ mole)\,(17\ g/mole)$$

$$\#\,g = 1.7\ g\ NH_3$$

Example 12.8: Determine the number of grams of iodine, I_2, necessary to form 41.2 grams of PI_3 when I_2 reacts with phosphorus, P, according to the equation:

$$P + I_2 \rightarrow PI_3$$

Step 1: Balance the equation.

$$2 P + 3I_2 \rightarrow 2 PI_3$$

Step 2: Determine the number of moles of PI_3.

$$\# \text{moles} = \frac{\# g}{GMW}$$

$$= \frac{41.2 \text{ g}}{412 \text{ g/mole}}$$

$$\# \text{moles} = 0.10 \text{ moles } PI_3$$

Step 3: Determine the number of moles of I_2 that will react with 0.10 moles PI_3.

$$\frac{x \text{ moles } I_2}{3 \text{ moles } I_2} = \frac{0.10 \text{ moles } PI_3}{2 \text{ moles } PI_3}$$

$$2x = 0.30$$

$$x = 0.15 \text{ moles } I_2$$

Step 4: Determine the number of grams of I_2 present.

$$\# \text{moles} = \frac{\# g}{GMW}$$

$$\# g = (\# \text{moles}) \, (GMW)$$

$$= (0.15 \text{ moles}) \, (254 \text{ g/mole})$$

$$\# g = 38.1 \text{ grams } I_2$$

12.5 The Mole-Volume Calculation

This type of calculation is based on the fact that 1 mole of an ideal gas at standard conditions occupies 22.4 liters. When working with a balanced chemical equation, the number of moles of a gas can be changed to volume by working with the above relationship. If the reaction is not at standard conditions, the volume of the gas must be changed to standard conditions before continuing.

Example 12.9: What volume of O_2 at standard conditions can be obtained when 5 moles of Ag_2O decomposes according to the equation:

$$Ag_2O \rightarrow Ag + O_2\,(g)$$

Step 1: Balance the equation.

$$2\,Ag_2O \rightarrow 4\,Ag + O_2\,(g)$$

Step 2: Determine the number of moles of O_2 formed when 5 moles Ag_2O decomposes.

$$\frac{5 \text{ moles } Ag_2O}{2 \text{ moles } Ag_2O} = \frac{x \text{ moles } O_2}{1 \text{ mole } O_2}$$

$$2x = 5 \text{ moles}$$

$$x = 2.5 \text{ moles } O_2$$

Step 3: Determine the volume of O_2 by using the relationship that 1 mole of gas = 22.4 liters at standard conditions.

$$\frac{2.5 \text{ moles } O_2}{1 \text{ mole } O_2} = \frac{x \text{ liters}}{22.4 \text{ liters}}$$

$$x = 56 \text{ liters } O_2$$

Example 12.10: Calculate the volume of O_2 necessary to react with 6 gram-atoms of iron, Fe, according to the following equation:

$$Fe + O_2 \rightarrow Fe_2O_3 \text{ at standard conditions.}$$

Step 1: Balance the equation.

$$4\,Fe + 3\,O_2 \rightarrow 2\,Fe_2O_3$$

Step 2: Determine the number of moles of O_2 necessary when 6 gram-atoms of Fe reacts.

$$\frac{6 \text{ gram-atoms Fe}}{4 \text{ gram-atoms Fe}} = \frac{x \text{ moles } O_2}{3 \text{ moles } O_2}$$

$$4x = 18 \text{ moles}$$

$$x = 4.5 \text{ moles } O_2$$

Step 3: Determine the volume of O_2 at standard conditions.

$$\frac{4.5 \text{ moles } O_2}{1 \text{ mole } O_2} = \frac{x \text{ liters } O_2}{22.4 \text{ liters } O_2}$$

$$x = 100.8 \text{ liters } O_2$$

$$x = 101 \text{ liters } O_2$$

12.6 The Mass-Volume Calculation

The mass-volume calculation involves one more step than the mole-volume calculation. That is, you must change the mass expression to a molar expression before determining the volume. The molar volume of an ideal gas is still used in the calculation.

Example 12.11: Determine the volume of H_2 produced when 180 grams of H_2O reacts according to the following equation at standard conditions.

$$Na + H_2O \rightarrow NaOH + H_2(g)$$

Step 1: Balance the equation.

$$2\,Na + 2\,H_2O \rightarrow 2\,NaOH + H_2$$

Step 2: Determine the number of moles of H_2O present.

$$\# \, moles = \frac{\# \, g}{GMW}$$

$$= \frac{180\,g}{18\,g/mole}$$

$$\# \, moles = 10\,moles\,H_2O$$

Step 3: Determine the number of moles of H_2 produced.

$$\frac{10\,moles\,H_2O}{2\,moles\,H_2O} = \frac{x\,moles\,H_2}{1\,mole\,H_2}$$

$$2x = 10\,moles\,H_2$$

$$x = 5\,moles\,H_2$$

Step 4: Determine the volume of H_2 at standard conditions.

$$\frac{5\,moles\,H_2}{1\,mole\,H_2} = \frac{x\,liters\,H_2}{22.4\,liters\,H_2}$$

$$x = (5)\,(22.4\,liters)$$

$$x = 112\,liters\,H_2$$

Example 12.12: Determine the number of grams of Al necessary to produce 67.2 liters of H_2 at standard conditions according to the following equation:

$$Al + HCl \rightarrow AlCl_3 + H_2$$

Step 1: Balance the equation.

$$2\,Al + 6\,HCl \rightarrow 2\,AlCl_3 + 3\,H_2$$

Step 2: Determine the number of moles of H_2.

$$\frac{x \text{ moles } H_2}{1 \text{ mole } H_2} = \frac{67.2 \text{ liters } H_2}{22.4 \text{ liters } H_2}$$

$$x = \frac{67.2}{22.4}$$

$$x = 3 \text{ moles } H_2$$

Step 3: Determine the number of gram-atoms of Al.

$$\frac{x \text{ gram-atoms Al}}{2 \text{ gram-atoms Al}} = \frac{3 \text{ moles } H_2}{3 \text{ moles } H_2}$$

$$x = 2 \text{ gram-atoms Al}$$

Step 4: Determine the number of grams of Al.

$$\# \text{gram-atoms Al} = \frac{\# g}{GAW}$$

$$\# g = (\# \text{gram-atoms}) \, (GAW)$$

$$= (2 \text{ gram-atoms}) \, (27 \text{ g/g-a})$$

$$\# g = 54 \text{ grams Al}$$

12.7 The Volume-Volume Calculation

When working a volume-volume calculation, you are primarily concerned with gases. This type of problem can be worked two different ways. The first is to convert all volumes to moles and perform a mole-mole calculation. The second method is to use Avogadro's Hypothesis which states that gases under the same conditions of temperature and pressure contain the same number of molecules and thus occupy the same volume. Thus, by use of the second method, you can compare reacting volumes directly with a balanced equation. Example 12.13 will be worked by both methods to illustrate the differences in the two procedures.

Example 12.13: What volume of $CO(g)$ will react with 67.2 liters of $O_2 \, (g)$ according to the following equation at standard conditions?

$$CO(g) + O_2(g) \rightarrow CO_2(g)$$

Method A: Using Moles

Step 1: Balance the equation.

$$2CO(g) + O_2(g) \rightarrow 2CO_2(g)$$

Step 2: Determine the number of moles of O_2 (g).

$$\frac{x \text{ moles } O_2}{1 \text{ mole } O_2} = \frac{67.2 \text{ liters } O_2}{22.4 \text{ liters } O_2}$$

$$x = \frac{67.2}{22.4}$$

$$x = 3 \text{ moles } O_2$$

Step 3: Determine the volume of $CO(g)$.

$$\frac{6 \text{ moles CO}}{1 \text{ mole CO}} = \frac{x \text{ liters CO}}{22.4 \text{ liters CO}}$$

$$x = (6)(22.4)$$

$$x = 134 \text{ liters of CO(g)}$$

Method B: Using Volume Relationship

Step 1: Balance the chemical equation.

$$2CO(g) + O_2 \rightarrow 2CO_2(g)$$

Step 2: Establish a proportion using the volume of O_2 given and the relationship from the balanced equation.

$$\frac{x \text{ volume CO}}{2 \text{ volumes CO}} = \frac{67.2 \text{ liters } O_2}{1 \text{ volume } O_2}$$

$$x = (2)(67.2 \text{ liters})$$

$$x = 134.4 \text{ liters CO(g)}$$

12.8 Concentration Calculations

A quick review of concentration calculations should involve molarity and normality. There are other methods of expressing concentration as discussed in Chapter 11, but molarity and normality are the two of major importance to the beginning student.

Molarity was defined as the number of moles of solute per liter of solution. In equation form, M = Molarity = $\frac{\text{\# moles of solute}}{\text{liters of solution}}$ Using this relationship, all molarity calculations can be performed.

Example 12.14: Calculate the molarity of a solution formed by adding 5.6 grams of KOH to enough water to make 500 ml of solution.

Step 1: Determine the number of moles of solute.

$$\# \, moles = \frac{\# \, g}{GMW}$$

$$= \frac{5.6 \, g}{56 \, g/mole}$$

$$\# \, moles = 0.10 \, moles \, KOH$$

Step 2: Determine the molarity.

$$M = \frac{\# \, moles \, of \, solute}{liters \, of \, solution}$$

$$M = \frac{0.10 \, moles}{0.50 \, liters}$$

$$M = 0.20 \, moles/liter$$

Example 12.15: 12.3 liters of HCl(*g*) at 27°C and 1.0 atm is totally dissolved in enough water so that the resultant solution has a volume of 6.2 liters. What is the molarity of the solution?

Step 1: Using the Equation of State for an ideal gas, determine the number of moles of HCl present.

$$PV = nRT$$

$$n = \frac{PV}{RT} = \frac{(1.0 \, atm) \, (12.3 \, liters)}{(.082 \, liter\text{-}atm/°mole) \, (300°K)}$$

$$n = 0.50 \, moles \, HCl \, (g)$$

Step 2: Determine the molarity.

$$M = \frac{\# \, moles \, solute}{liters \, of \, solution}$$

$$M = \frac{0.50 \, moles}{6.2 \, liters} = 0.081 \, M$$

Example 12.16: Given a 2.0 M solution of $AgNO_3$, determine the number of grams of $AgNO_3$ present in 0.70 liters of solution.

Step 1: Determine the number of moles of solute.

$$M = \frac{\# \, moles \, of \, solute}{liter \, of \, solution}$$

$$\# \, moles = (M) \, (liters)$$

$$= (2.0 \, M) \, (0.70 \, liters)$$

$$\# \, moles = 1.4 \, moles \, AgNO_3$$

Step 2: Determine the number of grams of $AgNO_3$.

$$\# \text{moles} = \frac{\#g}{MW}$$

$$\#g \quad = (\#\text{moles})\ (MW)$$

$$= (1.4\ \text{moles})\ (170\ \text{g/mole})$$

$$\#g \quad = 238\ \text{grams } AgNO_3 \approx 240\ \text{grams } AgNO_3$$

Normality calculations can cause problems because normality is based on the gram-equivalent weight and not the molecular weight of the reacting substance. To work a normality calculation, you must have a balanced chemical equation.

Normality is defined as the number of gram-equivalent weights of solute per liter of solution. In equation form:

$$N = \text{Normality} = \frac{\#\ \text{gram-equivalent wts}}{\text{liter of solution}}$$

It is also necessary to recall the expressions for determining the number of gram-equivalent weights and the equivalent weight.

If you are unsure of these relationships, refer back to Chapter 11 on normality calculations.

Example 12.17: Determine the normality of a $Ca(OH)_2$ solution in which 7.4 grams of $Ca(OH)_2$ has been dissolved in enough water to form 200 ml of solution. Assume complete dissociation of $Ca(OH)_2$.

Step 1: Determine the gram-equivalent weight.

$$GEW = \frac{GMW}{\#\ OH^-}$$

$$= \frac{74\ \text{g/mole}}{2}$$

$$GEW = 37\ \text{grams}$$

Step 2: Determine the number of gram-equivalent weights present in 7.4 grams of $Ca(OH)_2$.

$$\#\ GEW = \frac{\#g}{GEW}$$

$$= \frac{7.4\ \text{grams}}{37\ \text{grams}}$$

$$\#\ GEW = 0.20\ GEW\text{'s}$$

Step 3: Determine the normality.

$$N = \frac{\text{\# GEW of solute}}{\text{liter of solution}}$$

$$N = \frac{0.2 \text{ GEW's}}{0.2 \text{ liters}}$$

$$N = 1 \text{ GEW/liter}$$

Example 12.18: Calculate the normality of a solution of $AlCl_3$ if 1.34 grams of $AlCl_3$ is dissolved in enough water to make 200 ml of solution. Assume in a subsequent reaction that aluminum is converted from Al^{3+} to Al^0.

Step 1: Determine the gram-equivalent weight of $AlCl_3$.

$$GEW = \frac{GMW}{\# e^-}$$

$$= \frac{134 \text{ g/mole}}{3 \, e^-}$$

$$= 44.7 \text{ grams}$$

Step 2: Determine the number of gram-equivalent weights present.

$$\# GEW = \frac{\# g}{GEW}$$

$$= \frac{1.34 \text{ grams}}{44.7 \text{ grams}}$$

$$\# GEW = 0.0300$$

Step 3: Determine the normality.

$$N = \frac{\# GEW}{\text{liter}}$$

$$N = \frac{0.0300 \text{ GEW}}{0.200 \text{ liters}}$$

$$N = 0.150 \text{ GEW/liter}$$

12.9 pH Calculations

The pH of a solution is a measure of the hydrogen ion or hydronium ion concentration of the solution. A range has been established to indicate the concentration of hydrogen ions in a solution. Such a range is called a pH scale. On one end of this range are the acids or in other words, those reagents with a high hydrogen ion concentration. At the other end of the pH range are the bases or those reagents with a low hydrogen ion concentration and subsequently a high hydroxide ion concentration. The midpoint of the pH scale is called the neutral point and is characterized by a reagent which has an equal concentration of hydrogen and hydroxide ions. Such a reagent is water. Theoretically, the pH scale can vary as far as solubility and dissociation of the reagents will allow. Thus, both positive and negative pH's are possible. In practice, however, it has become customary to limit the pH scale to values between 0 and 14. The neutral point on this scale is 7. A substance is said to be acidic if it has a pH reading anywhere below 7. Accordingly, a substance is said to be basic if it has a pH reading above 7.

To calculate the pH of a solution, the hydrogen ion concentration of the solution must be available. pH is defined as the negative logarithm of the hydrogen ion concentration. In equation form: $pH = -\log [H^+]$.

Example 12.19: Determine the pH of a solution which has a hydrogen ion concentration of 1.0×10^{-3} M.

   Solution:

$$pH = -\log [H^+]$$
$$pH = -\log [1.0 \times 10^{-3}]$$
$$pH = -[\log 1.0 + \log 10^{-3}]$$
$$pH = -[0.0 - 3.0]$$
$$pH = -[-3.0]$$
$$pH = 3.0$$

Note: The log of 1.0 is zero and since we are working with base 10 numbers, the log of a power of 10 is merely the exponent. Thus, the log of 10^{-3} is the exponent -3. For more elaborate calculations, it is necessary to refer to log tables or a slide rule for the proper values. There is a table of logarithms in Appendix G of this text.

Example 12.20: Determine the pH for a solution which has a hydrogen ion concentration of 5.0 N 10^{-4} M.

   Solution:

$$pH = -\log [H^+] \qquad\qquad pH = -[0.7 - 4.0]$$
$$pH = -\log [5.0 \times 10^{-4} M] \qquad\qquad pH = -[-3.3]$$
$$pH = -[\log 5.0 + \log 10^{-4}] \qquad\qquad pH = 3.3$$

Note: The log of 5.0 is found in a log table to be 0.6990 or 0.70.

This value is then substituted into the equation and added algebraically.

Glossary

Molality—A measure of concentration. Molality *(m)* is equal to the number of moles of solute per kilogram of solvent.

Molarity—A measure of concentration. Molarity *(M)* is equal to the number of moles of solute per liter of solution.

Normality—A measure of concentration. Normality *(N)* is equal to the number of gram-equivalent weights of solute per liter of solution.

pH—A measure of the hydrogen ion concentration of a reagent. *pH* is equal to the negative logarithm of the hydrogen ion concentration. In equation form: $pH = -\log [H^+]$.

Exercises

1. Determine the number of moles of the reagent identified in each of the following substances.

   a. 10 grams of $CaCO_3$ (*s*)

   b. 200 ml of 2 *M* HCl solution

   c. 20 ml of 3 *N* H_2SO_4 solution

   d. 440 grams of CO_2 (*s*)

2. Determine the number of grams of the reagent identified in each of the following substances.

   a. 3 moles $MgCl_2$ (*s*)   c. 5 ml of 16 *M* HNO_3

   b. 50 ml of 1 *M* H_2SO_4 d. 100 ml of 0.02 *N* $CA(OH)_2$

3. According to the reaction: $2BaO_2 \rightarrow 2BaO + O_2$, how many moles of BaO will be produced when 3.0 moles of BaO_2 decompose?

4. According to the equation: $2Al + 6HCl \rightarrow 2AlCl_3 + 3H_2$, determine the number of gram-atoms of Al necessary to react with 5.0 moles of HCl.

5. According to the equation: $Ca + 2HCl \rightarrow CaCl_2 + H_2$, calculate the number of grams of Ca necessary to react with 2.0 moles of HCl.

6. According to the equation: $2S + 3O_2 \rightarrow 2SO_3$, determine the number of gram-atoms of S needed to react with 3.2 grams of O_2.

7. According to the reaction: $2Al + 6HCl \rightarrow 2AlCl_3 + 3H_2$, calculate the mass of H_2 produced when 270 grams of Al reacts.

8. Consider the reaction $2Na + 2H_2O \rightarrow 2NaOH + H_2$. If 4.6 grams of sodium reacts, what mass of NaOH would be produced?

9. Consider the reaction $2Na + 2HCl \rightarrow 2NaCl + H_2$. If 3.65 grams of HCl reacts, what mass of NaCl would be produced? What volume of H_2 would be produced at standard conditions?

10. In the reaction $2Al + 6HCl \rightarrow 2AlCl_3 + 3H_2$, what volume of H_2 would be produced if 4 moles of HCl reacts with sufficient Al at standard conditions?

11. Given 13.5 grams of Al, calculate the volume of H_2 produced in the reaction $2Al + 3H_2SO_4 \rightarrow Al_2(SO_4)_3 + 3H_2$ at standard conditions?

12. According to the reaction: $Mg + 2HCl \rightarrow MgCl_2 + H_2$, what mass of Mg would be required to produce 6.72 liters of H_2 at standard conditions?

13. Given the reaction $N_2(g) + 3H_2(g) \rightarrow 2NH_3(g)$, determine the volume of $NH_3(g)$ produced when 10 liters of $H_2(g)$ reacts to completion with excess $N_2(g)$.

14. Consider the equation $2H_2(g) + O_2(g) \rightarrow 2H_2O(g)$, what volume of $H_2(g)$ would be required to produce 12 liters of $H_2O(g)$? Assume sufficient O_2.

15. Determine the molarity of a solution formed by dissolving 9.5 grams of $MgCl_2$ in enough water to make 750 ml of solution.

16. Determine the molarity of the following solutions:
    a. 17 grams $AgNO_3$ in 400 ml of solution
    b. 2 moles of KOH in 5 liters of solution
    c. 3.65 grams of HCl in 200 ml of solution
    d. 100 grams Na_2CO_3 in 500 ml of solution

17. Determine the number of grams of $Mg(NO_3)_2$ present in 50 ml of a 3.5M solution of $Mg(NO_3)_2$.

18. Given the reaction: $Mg + 2HCl \rightarrow MgCl_2 + H_2$, determine the gram-equivalent mass of Mg.

19. Calculate the normality of a solution formed by dissolving 1.5×10^{-3} grams of $Ba(OH)_2$ in 800 ml of solution.

20. Calculate the pH of each of the following: (Assume 100% ionization).
    a. A solution with a H^+ concentration of 1.0×10^{-9} M
    b. A solution with a H^+ concentration of 2.0×10^{-3} M
    c. A 1.0×10^{-5} M solution of HCl
    d. A 3.5×10^{-2} M solution of HCl

CHAPTER 13

Chemical Equilibrium

13.1 Reversible Reactions

Thus far anytime a chemical equation has been used to describe a chemical reaction or chemical change, it has been discussed as a group of reactants being converted to a group of products. In most cases, this is an oversimplification. That is, in many chemical reactions the reverse operation also occurs. In other words, the products in turn react to form the initial reactants. When this change occurs, we say we have a *reversible reaction*. Consider the following example of a reversible process.

a. H_2O (solid) $\rightarrow$ H_2O (liquid)

b. H_2O (solid) $\rightarrow$ H_2O (liquid)

Equation (a) is an equation illustrating the melting of water. As written, the solid water (ice), is heated and converted to liquid water. Equation (b) illustrates the freezing of water. It also is written as a reaction going to completion. Through the study of phase changes, we know that the above example is a reversible process. That is, it can react as in either case (a) or case (b) depending upon the conditions. Thus, this chemical system could be illustrated by a single equation:

$$H_2O \text{ (solid)} \underset{\text{cooling}}{\overset{\text{heat}}{\rightleftharpoons}} H_2O \text{ (liquid)}$$

The double arrow indicates a reversible reaction.

The terminology involved in reversible reactions is important. Consider the general equation:

$$A + B \rightleftharpoons C + D$$

As written, A and B are said to be the *reactants* and C and D are referred to as *products*. By definition, the chemical species written on the left are called reactants and those species on the right are called products. Since this is a reversible reaction, the products C and D can react to form the initial reactants A and B. Therefore, in terms of equilibrium, it is merely how the equation is written that determines which species are reactants and which species are products. Chemically, you can think of the reactants as the chemicals you begin with and the products as the substances which you end up with in the initial or forward reaction.

The *forward reaction* is the reaction which proceeds from left to right as written. The *reverse reaction* is the reaction which proceeds from right to left as written. This is by convention. It is important to remember the terminology as it is used in equilibrium expressions; you must be able to recognize reactants and products.

13.2 Mass Action Expressions

In a reversible reaction, the forward reaction and the reverse reaction occur simultaneously. However, the degree to which the forward reaction and the reverse reaction occur is not the same. That is, there usually is a tendency for the reaction to proceed in one direction more than in the other direction. The extent of the reaction can be regulated by adjusting the experimental conditions to obtain the desired effect. The process of regulating or adjusting equilibrium will be discussed in Section 13.4. For now, remember that the forward and reverse reaction rates are not necessarily occurring at the same rate.

Consider the general equation:

$$aA + bB \rightleftharpoons cC + dD$$

To interpret this equation in words, you would say, "a moles of reactant A reacts with b moles of reactant B to form c moles of product C and d moles of product D." Through experimentation, a quantitative relationship based on this generalized equation has been developed. The quantitative relationship is:

$$\frac{[C]^c \times [D]^d}{[A]^a \times [B]^b} = \text{(At constant temperature)}$$

This relationship is known as the *Law of Mass Action.* The constant obtained in the above expression is known as the equilibrium constant and given the symbol K. The square brackets indicate concentration of the given species expressed in moles/liter.

13.3 Le Chatelier's Principle

In the previous section, it was mentioned that a chemist can adjust the experimental conditions of a reaction to favor either the forward or the reverse reaction. Related to this idea is *Le Chatelier's Principle* which states: if a stress is applied to a system at equilibrium, the system will tend to compensate for the applied stress and establish equilibrium under the new set of conditions.

The stresses that a system at equilibrium might encounter could be a change in concentration of one or more of the species in the system, a change in temperature, a change in pressure or volume, or the addition of a catalyst. Let us consider each of these stresses on a system which is at equilibrium.

For reference, the equilibrium reaction between nitrogen gas, N_2, and hydrogen gas, H_2, will be studied. The reaction is $N_2 + H_2 \rightleftharpoons NH_3 + $ kcal. As in any system, the first step should be to balance the equation. Balanced, the equation reads: $N_2 + 3H_2 \rightleftharpoons 2NH_3 + 23$ kcal. The first stress we should study is the effect of concentration.

Case 1: At constant temperature the concentration of H_2 is increased. The following changes occur:

a. The rate of the forward reaction would increase.
b. The concentration of N_2 would decrease as it is needed to react with the additional H_2.
c. The concentration of NH_3 would increase.
d. After a period of time, the rate of the forward reaction and the reverse reaction are equal.
e. The system has established a new equilibrium with the concentration of N_2 less than the original equilibrium and the concentration of NH_3 greater than the original equilibrium.
f. Conclusion: The net reaction has occurred to the right.

Case 2: The concentration of NH_3 is increased. The following changes occur:

a. The rate of the reverse reaction increases.
b. The concentration of N_2 and H_2 would increase.
c. The concentration of NH_3 would decrease from its increased value.

d. After a period of time, the rates of the forward and reverse reactions are equal.

e. The system has established a new equilibrium with the concentration of N_2 and H_2 greater than the original equilibrium and the concentration of NH_3 less.

f. Conclusion: The net reaction has occurred to the left.

Case 3: The temperature of the system is raised.

An increase in temperatures will favor the endothermic reaction or in other words, that reaction which requires heat or energy to occur. From the balanced equation, it is apparent that the reverse reaction requires energy. Thus, an increase in temperature will favor the reverse reaction. The following changes occur:

a. The concentration of NH_3 decreases.
b. The concentration of H_2 and N_2 increases.
c. The rate of the reverse reaction increases.
d. After a period of time, the two rates of reaction are again equal.
e. The system has established a new equilibrium with the concentration of N_2 and H_2 greater than in the original equilibrium, and the concentration of NH_3 has decreased from its original value.
f. Conclusion: The net reaction has occurred to the left.

A decrease in temperature would have the opposite effect. That is, a decrease in temperature favors the exothermic reaction or the reaction which liberates heat. In this case, a drop in temperature would favor the forward reaction and the equilibrium would shift to the right. The key to a temperature determination is to notice where the energy term is written. This will immediately tell you which is the endothermic and exothermic reactions.

Case 4: The pressure of the system is increased. (The volume of the system is decreased.) Notice that an increase in pressure may be treated the same as a decrease in volume. The key to pressure and volume determination is the number of molecules or moles of gas present on each side of the equation. This is because the greater the number of moles of gas, the greater the pressure. When the pressure of the system is raised, the system will adjust so as to try to obtain a lower pressure at equilibrium. To accomplish this result, the equilibrium must shift to the side with the fewer molecules of gas when the pressure is increased.

When the pressure is decreased, the equilibrium will shift to the side with the greater number of gaseous molecules.

For Case 4, when the pressure is increased, the following changes occur:

a. The rate of the forward reaction increases.
b. The concentration of N_2 and H_2 decreases.
c. The concentration of NH_3 increases.
d. After a period of time, the rates of the forward and reverse reactions are equal.
e. Conclusion: The net reaction has occurred to the right.

Case 5: The addition of a catalyst.

A catalyst does not affect the equilibrium of a chemical system. It does affect the rate of reaction for both the forward and reverse reactions. However, it affects them equally and, therefore, there is no change in the equilibrium position.

Example 13.1: Given the following chemical system at equilibrium:

$$2CO_2 + 135 \text{ kcal} \rightleftharpoons 2CO + O_2$$

State the effect of each of the following stresses:

a. Increase in pressure.
b. Decrease in temperature.
c. Increase the concentration of O_2.
d. Add a catalyst.

Solution:

a. An increase in pressure will cause the net reaction to occur to the side with the fewer number of molecules; in this case, to the left.
b. A decrease in temperature will favor the exothermic reaction. In this reaction, the reverse reaction is exothermic. Therefore, the net reaction occurs to the left.
c. An increase in the concentration of O_2 will cause a decrease in the concentration of CO and an increase in the concentration of CO_2. The net reaction will occur to the left.
d. A catalyst will affect both rates of reaction equally, causing no change in the point of equilibrium.

13.4 Equilibrium Constants

To this point, a qualitative view of equilibrium has been made. We are now going to consider some simple quantitative relationships involving the equilibrium constants.

The equilibrium constant has a variety of forms depending on the type of chemical system involved. A capital K is used to refer to a general equilibrium expression. K_A refers to the equilibrium expression for a weak acid and K_B refers to the equilibrium expression of a weak base. K_w is a special equilibrium expression relating to the dissociation of water. K_{sp} is used to calculate the solubility equilibrium for various salts.

As will be seen in the following sections, no matter what equilibrium expression is used, the mathematical set-up and calculation remains the same. This is because the mass action expression for all equilibrium calculations assumes the same format.

13.5 Gas Equilibrium

In the next three sections, we are going to study the quantitative aspects of various chemical equilibria. One of the simplest is a pure gaseous system. That is, a system composed of only gases.

The equilibrium constant for a gaseous system can be calculated two ways depending on the given or known information. If the concentration of the gases is given in terms of moles per liter, a standard equilibrium set-up is used. If, however, the concentration is not given but rather the partial pressure of each gas is given, you can use these pressure expressions to calculate the equilibrium constant of the system. The following examples illustrate both types of equilibrium calculations.

Example 13.2: Consider the chemical system represented by the equation:

$$H_2(g) + Br_2(g) \rightleftharpoons 2HBr(g) \text{ (Constant temperature)}$$

After equilibrium is attained, the concentration of $H_2(g)$ is 0.4 moles/liter, the concentration of $Br_2(g)$ is 0.1 moles/liter, and the concentration of HBr(g) is 0.02 moles/liter. Calculate the equilibrium constant for the system.

Solution:

Step 1: Write the mass action expression.

$$K = \frac{[\text{HBr}]^2}{[\text{H}_2][\text{Br}_2]}$$

Step 2: Substitute concentrations into expression.

$$K = \frac{[0.02]^2}{[0.4] \quad [0.1]}$$

$$K = \frac{0.0004}{0.04}$$

$$K = 1 \times 10^{-2}$$

Notice that K is a constant and has no units assigned to it.

Example 13.3: Consider the chemical system represented by the equation:

$$\text{H}_2(g) + \text{I}_2(g) \rightleftharpoons 2\text{HI}(g)$$

After equilibrium is attained, the partial pressure of the gases are as follows:

$$P_{\text{H}_2} = 0.50 \text{ atm}$$

$$P_{\text{I}_2} = 1.5 \text{ atm}$$

$$P_{\text{HI}} = 1.1 \text{ atm}$$

Calculate the equilibrium constant.

Solution: In this case, the concentration of the gases is given in terms of pressure. Therefore, the equilibrium constant obtained is written as K_p to indicate the calculation is based on partial pressure of the gases.

Step 1: Write the Mass-Action Expression

$$K_p = \frac{(P_{\text{HI}})^2}{P_{\text{H}_2} \cdot P_{\text{I}_2}}$$

Step 2: Substitute the partial pressures into the expression.

$$K_p = \frac{(1.1 \text{ atm})^2}{(0.50)(1.5)} =$$

$$K_p = 1.6$$

In terms of mathematics, it makes no difference whether the concentration is expressed in moles per liter or atmospheres. However, there may be a difference in the equilibrium constant obtained using partial pressures as compared to moles per liter. The difference in the equilibrium constant can be related to the Equation of State for an ideal gas. Remember that $PV = nRT$ and therefore, the pressure, P, equals $\frac{nRT}{V}$. Since n/v is a given concentration in terms of moles/liter, the relationship for a gas A would become $P_A = [A]\,RT$.

13.6 Gas and Solid Equilibrium

The key to a gas-solid calculation is to realize that the solid is of invariant concentration and is not included in the Mass Action Expression. In general, anytime a solid exists in a gaseous or liquid equilibrated system, the solid is not included in the equilibrium expression because its concentration does not change. All other calculations are carried out in the normal manner as illustrated in Example 13.4.

Example 13.4: The following system is allowed to attain equilibrium at 27°C. The pressure of $NH_3\,(g)$ and $H_2S(g)$ was found to be 0.14 atm for each. Calculate the equilibrium constant:

$$NH_4HS(s) \rightleftharpoons NH_3\,(g) + H_2S(g)$$

Solution:

Step 1: Write the Mass Action Expression.

$$K_p = \frac{P_{NH_3} \cdot P_{H_2S}}{[NH_4HS(s)]}$$

The concentration of $NH_4\,HS(s)$ is eliminated from the expression because it is constant. The expression then becomes:

$$K_p = P_{NH_3} \cdot P_{H_2S}$$
$$K_p = (0.14\ \text{atm})(0.14\ \text{atm})$$
$$K_p = 0.0196 =$$
$$K_p = 0.020$$

Now let us solve this same example in terms of K_c. All concentrations must be expressed in moles/liter.

Solution:

Step 1: Write the Mass Action Expression:

$$K_c = \frac{[NH_3]\,[H_2S]}{[NH_4HS(s)]}$$

Since the concentration of $NH_4HS(s)$ is constant, the term is removed from the expression. Therefore:

$$K_c = [NH_3][H_2S]$$

Step 2: For a given gas, the relationship between partial pressure and the concentration in terms of moles per liter is $P_A = [A]$ RT or $[A] = \dfrac{P_A}{RT}$ Substitute this relationship into the above equation for each gas.

$$K_c = \frac{P_{NH_3}}{RT} \cdot \frac{P_{H_2S}}{RT}$$

$$K_c = \frac{.14\text{ atm}}{(.082\text{ l-atm }/^\circ\text{ mole})(300^\circ K)} \cdot \frac{.14\text{ atm}}{(.082\text{ l-atm }/^\circ\text{ mole})(300^\circ K)}$$

$$= \frac{(.14)^2}{(.082 \cdot 300)^2} = \frac{.0196}{605} = 3.2 \times 10^{-5}$$

Example 13.5: The following system is allowed to attain equilibrium at 127° C. The pressure of CO_2 (g) was found to be 0.22 atm. Calculate K_p and K_c. $CaCO_3(s) \rightleftharpoons CaO(s) + CO_2(g)$.

Solution:

Step 1: Write the Mass Action Expression.

$$K_p = P_{CO_2}$$

Note: Both $CaCO_3(s)$ and $CaO(s)$ are solids and thus their concentrations are constant and not included in the expression. Therefore:

$$K_p = P_{CO_2} = 0.22$$

Step 2: Calculate K_c.

$$K_c = [CO_2] \text{ and } [CO_2] = \frac{P_{CO_2}}{RT}$$

$$K_c = \frac{0.22\text{ atm}}{(0.082\text{ 1-atm}/^\circ\text{mole})(300^\circ K)}$$

$$K_c = 0.0089 = 8.9 \times 10^{-3}$$

13.7 Ionic Equilibrium

In Chapter 5 we learned that when certain substances are put in solution, their molecules dissociate to form charged particles called ions. It must be emphasized that not all molecules ionize or form ions to the same extent. That is, a substance such as HCl will for practical purposes, completely ionize to form H^+ ions and Cl^- ions while a substance such as acetic acid, CH_3COOH, will ionize about one percent.

In carrying out equilibrium calculations, the degree of ionization must be watched very carefully. The reason for this is that two basic assumptions are made when working equilibrium problems involving ions. The first assumption is that any substance which has a high degree of ionization is assumed to be completely ionized to form only ions with no molecular species remaining. An example would be the ionization of strong acids such as HCl and strong bases such as NaOH. The second assumption is that when a substance has a small degree of ionization, you assume negligible ionization with respect to the initial concentration; thus, only molecular species are present. An example of the second case would be a weak acid such as acetic acid.

Example 13.6: Calculate the equilibrium concentration of hydrogen ion and acetate ion in a system which initially contains 1.0 M acetic acid.

Solution:

Step 1: Write a balanced equation:

$$CH_3COOH(aq) \rightleftharpoons H^+(aq) + CH_3COO^-(aq)$$

Step 2: Write a Mass-Action Expression:

$$K_A = \frac{[H^+(aq)][CH_3COO^-(aq)]}{[CH_3COOH(aq)]}$$

Step 3: Substitute into the expression. The equilibrium constant for acetic acid is 1.8×10^{-5}. From the balanced equation, it is evident that for every molecule of CH_3COOH that dissociates, one H^+ ion and one CH_3COO^- ion is produced. Let x equal the amount of CH_3COOH which dissociates. A comparison of initial concentrations and equilibrium concentrations is shown in Table 13.1.

TABLE 13.1
A COMPARISON OF ACETIC ACID CONCENTRATIONS

Species	Initial Concentrations	Equilibrium Concentrations
CH_3COOH	1.0 M	$1.0 - x$
H^+	0.0 M (approx.)	x
$CH_3COO^=$	0.0 M	x

The equilibrium expression then becomes:

$$K_A = \frac{[H^+ (aq)] \; [CH_3COO^- (aq)]}{[CH_3COOH (aq)]}$$

$$1.8 \times 10^{-5} = \frac{[x] \; [x]}{[1.0M - x]}$$

Notice that in the above expression, the concentrations are *equilibrium* concentrations. Since there were essentially no H^+ ions or CH_3COO^- ions present, initially, the equilibrium concentration of these species will be only that which is produced from the dissociation of CH_3COOH. Also, the equilibrium concentration of CH_3COOH is found by taking the initial concentration and subtracting the amount which dissociates.

Step 4: We can now make an assumption which will simplify the calculation. Since acetic acid is a weak acid, we can assume the amount of acetic acid which dissociates is negligible with respect to the initial concentration. Therefore, the expression becomes:

$$1.8 \times 10^{-5} = \frac{[x] \; [x]}{[1.0 \; M]} \qquad \left(\begin{array}{l} \text{In the denominator, } x \text{ is} \\ \text{negligible compared to } 1.0 \; M. \end{array} \right)$$

Step 5: Solve the expression for x.

$$1.8 \times 10^{-5} = \frac{x^2}{1}$$

$$\sqrt{1.8 \times 10^{-5}} = \sqrt{x^2}$$

$$4.2 \times 10^{-3} M = x = [H^+] = [CH_3COO^-]$$

Answer:

$$[H^+] = 4.2 \times 10^{-3} \; M$$

$$[CH_3COO^-] = 4.2 \times 10^{-3} \; M$$

Note: For those interested in verifying this assumption, solve the expression in Step 3 using the quadratic formula or the mathematical process called completion of the square and compare your answer with the answer in Step 5.

Example 13.7: Calculate the equilibrium concentration of H^+ ion and Cl^- ion in a 2.0 M solution of HCl.

Solution:

Step 1: Write a balanced chemical equation for the system.

$$HCl(aq) \rightleftharpoons H^+(aq) + Cl^-(aq)$$

Step 2: Remember that HCl is a strong acid and consequently we assume 100% or complete ionization. Thus, all the HCl dissociates to form H^+ ions and Cl^- ions. From the balanced equation, we see there is a one-to-one ratio between HCl and the ions. Therefore, the equilibrium concentration of H^+ ion and Cl^- ion is 2.0 M. Table 13.2 summarizes the initial and equilibrium concentrations of HCl.

TABLE 13.2
INITIAL AND EQUILIBRIUM CONCENTRATIONS OF 2.0 M HCl

Species	Initial Concentration	Equilibrium Concentration (Assume complete dissociation)
HCl(aq)	2.0 M	0.0 M
H^+(aq)	0.0 M (approx.)	2.0 M
Cl^-(aq)	0.0 M (approx.)	2.0 M

The two previous examples are typical of the ionic equilibrium type. As one goes further in the study of equilibrium, it will become much more involved and require a more rigorous mathematical treatment. The basic concepts, however, do not change.

13.8 Solubility Equilibrium

Another form of equilibrium involves the ability of a salt to dissolve in a given amount of solvent to form a saturated solution. A measure of the ability of a substance to dissolve is known as the K_{sp} or solubility product. The solubility product is calculated in the same manner as all equilibrium constants except the denominator is one because the reactant is a solid. The larger the value of the K_{sp}, the more soluble the substance.

The concept of solubility is critical in qualitative analysis where substances are identified by selective precipitation of various salts. The precipitation is achieved by exceeding the solubility product of the given compound.

Since solubility is dependent upon experimental conditions, such as temperature, pressure, and solvent, it is essential that all conditions be listed when performing a solubility calculation.

Example 13.8: Calculate the solubility product of AgCl in water at 25°C if the solubility of AgCl is known to be 1.0 $\times$ 10^{-5} M.

Solution:

Step 1: Write a balanced chemical equation.

$$AgCl(s) \rightleftharpoons Ag^+(aq) + Cl^-(aq)$$

Step 2: Write K_{sp} expression.

$$K_{sp} = [Ag^+(aq)] \, [Cl^-(aq)]$$

Step 3: Substitute into the K_{sp} expression. From the example, the solubility of AgCl is 1.0×10^{-5} moles/liter. From the balanced equation, we see that each time one molecule of AgCl dissociates, one Ag^+ ion and one Cl^- ion are produced. Therefore, if 1.0×10^{-5} moles of AgCl dissociates, 1.0×10^{-5} moles Ag^+ and 1.0×10^{-5} Cl^- are produced. The K_{sp} expression becomes:

$$K_{sp} = [1.0 \times 10^{-5}][1.0 \times 10^{-5}]$$
$$K_{sp} = 1.0 \times 10^{-10}$$

Example 13.9: A solution at 25°C has a silver ion concentration of 1.5×10^{-4} M and a chloride ion concentration of 2.0×10^{-3} M. Will precipitation of AgCl(s) occur?

Solution:

Step 1: Write an expression to illustrate the *"ion product"* (*ip*). The ion product is the mathematical product of the various ion concentrations.

$$ip = [Ag^+(aq)] \, [Cl^-(aq)]$$

Step 2: Calculate the *ip* for the given concentrations.

$$ip = [1.5 \times 10^{-4}][2.0 \times 10^{-3}]$$
$$ip = 3.0 \times 10^{-7}$$

Step 3: Compare the *ip* value obtained with the accepted K_{sp} for AgCl at 25°C. If the *ip* value is greater, precipitation will occur; and if the *ip* value is less than the K_{sp}, no precipitation will occur.

$$K_{sp} = 1.0 \times 10^{-10} \text{ (from example 13.8)}$$
$$ip = 3.0 \times 10^{-7}$$

Therefore, precipitation *will* occur because 3.0×10^{-7} is larger than the K_{sp} value of 1.0×10^{-10}.

Example 13.10: Calculate the equilibrium concentration of Ca^{2+} ion and F^- ion in a saturated solution of CaF_2 at 22°C. The K_{sp} is 4.2×10^{-12}.

Solution:

Step 1: Write a balanced chemical equation.

$$CaF_2(s) \rightleftharpoons Ca^{2+}(aq) + 2 F^-(aq)$$

Step 2: Let x equal the solubility of CaF_2 in moles per liter and substitute into the K_{sp} expression.

$$Ksp = [Ca^{2+}(aq)] [F^-(aq)]^2$$
$$4.2 \times 10^{-12} = (x)(2x)^2$$

Note: The concentration of F^- ion is twice the solubility of CaF_2 because each molecule of CaF_2 dissociates to form *two* F^- ions. Refer to the balanced equation.

Step 3: Solve the K_{sp} expression.

$$4.2 \times 10^{-12} = (x)(2x)^2$$
$$4.2 \times 10^{-12} = x \cdot 4x^2$$
$$4.2 \times 10^{-12} = 4x^3$$
$$\frac{4.2 \times 10^{-12}}{4} = x^3$$
$$\sqrt[3]{1.05 \times 10^{-12}} = \sqrt[3]{x^3}$$
$$1.0 \times 10^{-4} = x$$

Step 4: Determine the concentration of Ca^{2+} and F^-.

$$[Ca^{2+}] = x = \underline{1.0 \times 10^{-4} M}$$
$$[F^-] = 2x = \underline{2.0 \times 10^{-4} M}$$

Glossary

Dissociation—The chemical process of separating or disconnecting the component parts of a molecule. Usually, the component parts are ions.

Equilibrium—A state of dynamic balance between the forward and reverse reactions of a chemical system in which the rates of the reactions are equal.

Ionization—The process of separating a molecule into its component ions. The process of forming charged particles called ions.

Law of Mass Action—An expression which relates the concentrations of the reactants and the products in a chemical system at equilibrium.

For the general equation $aA + bB = cC + dD$, the expression is written as

$$K = \frac{[C]^c \, [D]^d}{[A]^a \, [B]^b}.$$

Le Chatelier's Principle—If a stress is applied to a system in dynamic equilibrium, the system will tend to compensate for the applied stress and establish a new point of equilibrium.

Reversible Reaction—A chemical reaction which procedes in both the forward and reverse directions as written.

Solubility Product Constant—The product of the concentrations of the ions of a substance in a saturated solution of the substance.

Exercises

1. Formulate Mass Action expressions in terms of (a) partial pressures and (b) concentrations for the following equilibria:

   a. $H_2(g) + I_2(g) \rightleftharpoons 2HI(g)$

   b. $2Br_2(g) + 2H_2O(g) \rightleftharpoons 4HBr(g) + O_2(g)$

2. For each of the following equilibria predict the affect of (a) an increase in pressure, (b) a decrease in temperature and (c) a catalyst is added.

   a. $25 \text{ kcal} + N_2(g) + 3H_2(g) \rightleftharpoons 2NH_3(g)$

   b. $N_2(g) + O_2(g) + \text{Heat} \rightleftharpoons 2NO(g)$

   c. $C(s) + H_2O(g) + \text{Heat} \rightleftharpoons CO(g) + H_2(g)$

   d. $N_2O_4(g) \rightleftharpoons 2NO_2(g) + \text{Heat}$

3. Write equilibrium constant expressions for the following reactions:

   a. $H_2(g) + Cl_2(g) \rightleftharpoons 2HCl(g)$

   b. $2Ag^+(aq) + NO_3^-(aq) + Cu^0(s) \rightleftharpoons Cu^{2+}(aq) + NO_3^-(aq) + 2Ag^0(s)$

   c. $H_2SO_4(aq) \rightleftharpoons H^+(aq) + HSO_4^-(aq)$

   d. $NH_4OH(aq) \rightleftharpoons NH_4^+(aq) + OH^-(aq)$

   e. $Al_2(SO_4)_3(s) \rightleftharpoons 2Al^{3+}(aq) + 3SO_4^{2-}(aq)$

4. Write the equations for the three-step dissociation of phosphoric acid, H_3PO_4. Write an equilibrium constant expression for each reaction.

5. Classify each of the following systems as acidic, basic, or neutral when equilibrium is achieved.

   a. 100ml $6\,M$ HCl and 100ml $6\,M$ NH_4OH

b.   100ml $2\,M$ HNO_3 and 100ml $2\,M$ NaOH

c.   100ml $1\,M$ CH_3COOH and 100ml $1\,M$ KOH

d.   500ml $6\,M$ HCl and 100ml $6\,M$ NaOH

e.   200ml $12\,M$ HNO_3 and 400ml $6\,M$ KOH

6. Calculate the partial pressure of I_2 at equilibrium if the partial pressure of H_2 = 0.75 atm and the partial pressure of HI = 0.50 atm. The K_p = 6.25. The reaction:

$$H_2{}^+(g) + I_2(g) \rightleftharpoons 2HI(g)$$

7. Calculate the equilibrium constant of the following system knowing the equilibrium concentration of $N_2O_4(g)$ = 2.5 moles/liter and NO_2 = 1.2 moles/liter. The equation:

$$N_2O_4(g) \rightleftharpoons 2NO_2(g)$$

8. Calculate the equilibrium concentration of H^+ ion and $NO_3{}^-$ ion in a 6 M solution of HNO_3. (Assume total ionization.)

9. Calculate the solubility product of PbS in water at 25°C if the solubility of PbS is 2×10^{-14} moles/liter.

10. A solution at 25°C has a lead ion concentration of 1.2×10^{-20} moles/liter and a sulfide ion concentration of 2.0×10^{-18} moles/liter. Will precipitation of PbS(s) occur? Refer to Problem 9 for the K_{sp} of PbS.

CHAPTER 14

Oxidation-Reduction and Electrochemistry

14.1 Introduction

To this point in our study, chemical equations have been balanced by inspection or by a trial and error method. In more involved equations, it becomes necessary to have a systematic approach to the balancing of equations.

The assignment of oxidation states to chemical species and the subsequent balancing of equations by the method of electron transfer is one systematic approach to the balancing of equations. During the study of equation balancing, such areas as oxidation-reduction (redox) and electrochemistry are also encountered.

14.2 Determination of Oxidation State

Before attempting to balance redox equations, a student must be able to assign oxidation numbers to all the elements in a chemical equation. The following is an *arbitrary* set of rules for the assignment of oxidation states.

1. All elements in their free state have an oxidation state of 0. A free state is when the element has not combined with other elements other than itself. Some examples would be: Mg, Al, Fe, S, O_2, F_2.
2. Hydrogen has an oxidation state of +1 except in metallic hydrides where it is a −1.
3. Oxygen has an oxidation state of a −2 except in peroxides where it is a negative one.

4. All monoatomic ions have an oxidation state equal to the charge of the ion.
5. The algebraic sum of the oxidation states for a molecule must equal zero.
6. The algebraic sum of the oxidation states for an ion must equal the net charge on the ion.
7. It is possible to have fractional and zero oxidation states for elements in compounds.

Example 14.1: Determine the oxidation state for each element in the following substances:

$$\text{(a) } H_2O, \text{ (b) } SO_2, \text{ (c) } Na_3PO_4, \text{ (d) } NO_3{}^-$$

Solution:

a. H_2O — Water molecule. The sum of the oxidation states must equal zero for a neutral molecule. The oxidation state of hydrogen is $+1$ from rule 2 and the oxidation state of oxygen is -2 from rule 3.

b. SO_2 — Sulfur dioxide molecule. The sum of the oxidation states must equal zero. By rule 3, the oxygen is a -2. Since there are two oxygen atoms, the total negative charge is $(-2) + (-2) = -4$. Therefore, the sulfur is a $+4$ in order to have a neutral molecule.

c. Na_3PO_4 — Sodium phosphate molecule. The sum of the oxidation states must equal zero. Since sodium is in column one, it has an oxidation state of $+1$. Oxygen, by rule 3, has an oxidation state of -2. To obtain the oxidation state of phosphorus, set up an equation.

$$3(+1) + (P) + 4(-2) = 0. \quad P = +5$$

d. $NO_3{}^-$ — Nitrate ion. The sum of the oxidation states must equal the net charge on the ion. By rule 3, the oxygen is -2. To obtain the oxidation state for the nitrogen, set up an equation. $(N) + 3(-2) = -1$. $N = +5$. Note that the net charge on the nitrate ion is a negative one and therefore, the equation is set equal to a negative one.

14.3 Balancing Redox Equations

Just as there are rules for the determination of oxidation state, so there are rules for the balancing of redox equations. Of course, if you can balance the equation by inspection, you should do so as that is a much easier process. The steps for balancing redox equations are as follows:

Step 1: Assign oxidation numbers to each element in all compounds so as to determine which elements have been oxidized and which have been reduced.

Step 2: Write half-cell equations for the species that have been oxidized and reduced. Make sure the equations are balanced and that the electrons have been assigned to the side of your equation with the higher oxidation state.

Step 3: Combine the half-cell equations together algebraically so as to eliminate by addition the electrons from each side of the equation. This is done in the same manner as solving simultaneous equations.

Step 4: Insert the coefficients obtained from the half-cell equation back into the original equation.

Step 5: Balance the charge in the original equation.

Step 6: Balance the hydrogen and the oxygen in the original equation.

Note: To have an equation balanced:

1. The total charge on one side of the equation must equal the total charge on the other side of the equation.
2. There must be the same number of atoms of each element on each side of the equation.
3. The coefficients must be in least terms.

When a reaction occurs in an acidic solution, you can add hydrogen ions and water as necessary to balance the equation.

When a reaction occurs in a basic solution, you can add hydroxide ions and water as necessary to balance the equation.

Example 14.2: Balance the following redox equation:

$$Cu + NO_3^- + H^+ \rightleftharpoons Cu^{2+} + NO + H_2O$$

Solution:

Step 1: Assign oxidation states.

$$Cu^0 + (N^{5+}O^{2-}{}_3)^- + H^+ \rightleftharpoons Cu^{2+} + N^{2+}O^{2-} + H^+{}_2O^{2-}$$

Step 2: Determine which species have been oxidized and reduced. Write half-cell equations for each.

$$Cu^0 \rightarrow Cu^{2+} + 2e^-$$
$$4H^+ + NO_3^- + 3e^- \rightleftharpoons NO + 2H_2O$$

Note: The electrons are exchanged in such a way that the charge on each side of the half-cell equation is the same.

Step 3: Combine half-cell equations together algebraically so as to eliminate the electrons. In this case, we must multiply the top equation by 3 and the bottom equation by 2.

$$3 (Cu^0 \rightarrow Cu^{2+} + 2e^-)$$

$$\underline{2 (4H^+ + NO_3^- + 3e^- \rightarrow NO + 2H_2O)}$$

$$3Cu^0 + 8H^+ + 2NO_3^- + 6e^- \rightleftharpoons 3Cu^{2+} + 6e^- + 2NO + 4H_2O$$

Note: The electrons are eliminated because the same number appears on each side of the equation.

Step 4: Put the coefficients back in the original equation.

$$3Cu + 2NO_3^- + 8H^+ \rightarrow 3Cu^{2+} + 2NO + 4H_2O$$

Step 5: (a) Balance the charge. (b) Balance the hydrogen and the oxygen which will in turn balance the charge.

$$3Cu + 2NO_3^- + 8H^+ \rightarrow 3Cu^{2+} + 2NO + 4H_2O$$

The equation is balanced.

Example 14.3: Balance the following equation for the reaction occurring in an acidic solution.

$$NO_3^- + Cl^- \rightleftharpoons NO + Cl_2 \text{ (acid)}$$

Solution:

Step 1: Assign oxidation states.

$$(N^{5+} O^{2-}_3)^- + Cl^- \rightleftharpoons N^{2+} O^{2-} + Cl_2^0$$

Step 2: Write half-cell equations for the oxidized and reduced species.

$$NO_3^- + 4H^+ + 3e^- \rightarrow NO + 2H_2O$$
$$2Cl^- \rightarrow Cl_2^0 + 2e^-$$

Note: Balance the chlorine half-cell reaction by inspection. Write the *total* number of electrons required in the half-cell equation.

Step 3: Combine the half-cell equations algebraically.

$$2 (NO_3^- + 4H^+ + 3e^- \rightarrow NO + 2H_2O)$$

$$\underline{3(2Cl^- \rightarrow Cl_2^0 + 2e^-) \quad + 6e^-}$$

$$2NO_3^- + 6Cl^- + 8H^+ + 6e^- \rightleftharpoons 2NO + 4H_2O + 3Cl_2 + 6e^-$$

Step 4: Insert the coefficients back in the original equation.

$$8H^+ + 2NO_3^- + 6Cl^- \rightleftharpoons 2NO + 3Cl_2 + 4H_2O$$

Step 5: (a) Balance the charge. Since this reaction is occurring in an acid solution, you may add H^+ and H_2O as necessary. Notice the charge is balanced with a net charge of zero on each side. (b) Check the oxygen and hydrogen to see that they are also balanced. In this case, there are 8 hydrogens on each side and 6 oxygens. The equation is balanced:

$$8H^+ + 2NO_3^- + 6Cl^- \rightleftharpoons 2NO + 3Cl_2 + 4H_2O$$

Example 14.4: Balance the following equation for a reaction occurring in a basic solution.

$$MnO_4^- + NO_2^- \rightleftharpoons MnO_2 + NO_3^- \text{ (base)}$$

Solution:

Step 1: Assign oxidation states.

$$(Mn^{7+}O^{2-}_4)^- + N^{3+}O^{2-}_2 \rightleftharpoons Mn^{4+}O^{2-}_2 + (N^{5+}O^{2-}_3)^-$$

Step 2: Write half-cell equations for the oxidized and reduced species.

$$2H_2O + MnO_4^- + 3e^- \rightarrow MnO_2 + 4OH^-$$
$$+2OH^- + NO_2^- \rightarrow NO_3^- + 2e^- + H_2O$$

Step 3: Combine the half-cell equations algebraically.

$$2(2H_2O + MnO_4^- + 3e^- \rightarrow MnO_2 + 4OH^-)$$
$$3(2OH^- + NO_2^- \rightarrow NO_3^- + 2e^- + H_2O)$$

$$\overline{4H_2O + 2MnO_4^- + 6e^- + 6OH^- + 3NO_2^- \rightleftharpoons 2MnO_2 + 8OH^- + 3NO_2^- +}$$
$$6e^- + 3H_2O$$

Note: In this equation, there are water molecules and hydroxide ions on both sides of the equation. Add them algebraically so that each species will appear on but one side of the equation.

Step 4: Insert coefficients into the original equation.

$$H_2O + 2MnO_4^- + 3NO_2^- \rightleftharpoons 2MnO_2 + 2OH^- + 3NO_3^-$$

Step 5: (a) Balance the charge. The net charge is -5 on each side of the equation. (b) Balance the hydrogen and the oxygen. They are balanced with 14 oxygen and 2 hydrogen on each side of the equation.

Balanced equation:

$$H_2O + 2MnO_4^- + 3NO_2^- \rightleftharpoons 2MnO_2 + 2OH^- + 3NO_3^-$$

14.4 Electrochemical Cells

To study the relationships occurring in an electrochemical cell, it may be easiest to begin with a typical set-up of a cell and then expand the concepts obtained to electrochemical cells in general. A very common electrochemical cell is a chemical system composed of two beakers, a salt-bridge to complete the circuit, a strip of copper metal, a strip of silver metal, a 1.0 molar solution of silver nitrate and a 1.0 molar solution of copper(II) sulfate. Also included is a voltmeter and some wire for making proper connections.

The entire system is assembled as shown in Figure 14.1.

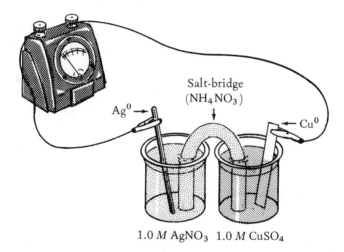

$1.0\ M\ AgNO_3$ $1.0\ M\ CuSO_4$

Figure 14.1. The Set-Up of a Typical Electrochemical Cell.

When all connections are made, the voltmeter will show a given amount of potential difference between the silver and copper electrodes. After a few minutes, measurable changes can be determined in the beakers. The copper electrode seems to be decreasing in mass and the silver electrode is showing an increase in mass. For convenience, let us assume that this change in mass is measured and the copper strip was found to have a mass loss of 0.64 grams and the silver strip was found to have a gain in mass of 2.16 grams.

At this point, let us calculate the number of gram-atoms of each metal lost or gained.

$$\text{\# of gram-atoms } Cu^0 \text{ lost} = \frac{\text{\# gram } Cu^0 \text{ lost}}{\text{AW Cu}}$$

$$= \frac{0.64 \text{ grams}}{63.5 \text{ g/g-A}}$$

$$= 0.01 \text{ gram-atoms } Cu^0$$

$$\text{\# of gram-atoms } Ag^0 \text{ gained} = \frac{\text{\# grams } Ag^0 \text{ gained}}{\text{AW } Ag^0}$$

$$= \frac{2.16 \text{ grams}}{108 \text{ g/g-A}}$$

$$= 0.02 \text{ gram-atoms } Ag^0$$

From these calculations, it is evident that the gram-atom ratio between Cu^0 and Ag^0 is a 1:2 ratio. That is for every gram-atom of Cu^0 that dissolved, two gram-atoms of Ag^0 were deposited on the silver strip.

By studying the beakers and the chemical species present, it is found that the Cu^0 metal that disappears is actually going into solution to form Cu^{2+} ions. This process can be represented by the equation:

$$Cu(s) = Cu^{2+}(aq) + 2e^-$$

In the other beaker, the silver is showing an increase in mass so it is apparent that Ag^- ions in solution are being changed to Ag^0 atoms on the metal strip. This reaction can be written as:

$$Ag^+(aq) + 1e^- \rightarrow Ag^0(s)$$

Since this is a completed chemical system with the wires and voltmeter supplying the necessary external circuit, it should also be apparent that the electrons liberated at the copper electrode pass through the external wires and migrate to the silver electrode where they combine with Ag^+ ions from the $AgNO_3$ solution to form silver atoms on the silver electrode. Notice how the two previous equations illustrate these processes.

In general, in all electrochemical cells there is at least one reaction that *donates* electrons in the system and also at least one reaction that *accepts* electrons in the system.

Such reactions are called *half-cell reactions*. Each electrochemical cell has two half-cell reactions. In the previous example, one half-cell reaction occurred in the beaker containing the copper strip and the copper sulfate solution. In symbol form, this is written $Cu^0/CuSO_4$. The other half-cell reaction occurred in the $AgNO_3/Ag^0$ beaker. The total reaction for the system would be obtained by adding the two half-cell equations together. That is:

$$\text{Multiply equation by (1) } (Cu^0(s) \rightarrow Cu^{2+}(aq) + 2e^-)$$
$$\text{Multiply equation by (2) } (Ag^+(aq) + 1e^- \rightarrow Ag^0(s))$$
$$\text{(total) } Cu(s) + 2Ag^+(aq) \rightleftharpoons Cu^{2+}(aq) + 2\,Ag(s)$$

14.5 Oxidation-Reduction Terminology

Proper terminology is essential when discussing oxidation-reduction reactions. The chemical species which loses electrons is said to be *oxidized*. In the previous example, the copper was oxidized. Another way of recognizing the oxidized species is to note which species has an increase in oxidation state during the chemical reaction. In the example, copper has gone from an oxidation state of zero in the free metal to a +2 in the Cu^{2+} ion.

The chemical species which gains electrons is said to be *reduced*. Silver was reduced in the example. Another method of remembering this is to realize that the chemical species reduced will always show a loss in oxidation state. In the example, silver went from +1 to zero. That is, the Ag^+ ion has an oxidation state of +1 and the free silver metal has an oxidation state of zero.

The *oxidizing agent* is the chemical species which is reduced. In the example, the Ag^+ ion is the oxidizing agent. The *reducing agent* is the species which is oxidized. In the example, the Cu^0 metal is the reducing agent.

In summary, oxidation-reduction reactions involves a change in oxidation number. The portion of the reaction involving the loss of electrons is called *oxidation* and that portion of the reaction involving the gain of electrons is called *reduction*. A shorthand notation for this type of reaction is said to be a *redox reaction* which is represented by a *redox equation*.

14.6 Half-Cell Equations—
Loss and Gain of Electrons

As mentioned in the previous sections, an electrochemical cell is composed of at least two half-cell reactions. At this time, let us consider these reactions in greater detail.

A student may wonder how we can predict the products in example one of this chapter. That is, how do we know that the copper metal changes to the cupric ion and the silver ion changes to metallic silver. One answer was mentioned earlier when we determined the mass of the electrodes showing that the silver electrode had an increase in mass and the copper electrode a reduction in mass.

Oxidation-reduction may also occur in a single beaker. For example, take a coil of copper and immerse it in a solution of silver nitrate. After a few minutes, you will notice gray crystals forming on the copper wire and also the solution is turning blue. Notice Figure 14.2.

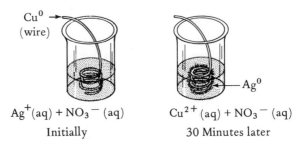

Cu^0 (wire)

$Ag^+(aq) + NO_3^-(aq)$

Initially

$Cu^{2+}(aq) + NO_3^-(aq)$

30 Minutes later

Ag^0

Figure 14.2. The Reaction of Copper and Silver Nitrate.

It is obvious from this demonstration that the copper has been oxidized to the blue cupric ion and the silver has been reduced to the gray metallic silver. As you will recall, the half-cell equations for this reaction were:

$$\text{(a) } 2\,(Ag^+(aq) + 1e^- \rightarrow Ag^0(s))$$
$$\text{(b) } 1\,(Cu^0(s) \rightarrow 2e^- + Cu^2\,(aq))$$
$$\overline{\text{(total) } Cu^0(s) + 2Ag^+(aq) \rightleftharpoons Cu^{2+}(aq) + 2Ag^0(s)}$$

As another example, consider the reaction between magnesium metal and sulfuric acid. Figure 14.3 illustrates this reaction in process.

Notice there are bubbles of gas surrounding the magnesium ribbon and as time progresses, the ribbon is completely dissolved to form magnesium ions. Since this reaction involves an acid, the likely gas is hydrogen which is formed from the hydrogen ions in the solution. The half-cell equation for this reaction would be:

$$\text{(oxidation)} \quad Mg^0(s) \rightarrow Mg^{2+}(aq) + 2e^-$$
$$\text{(reduction)} \quad 2H^+(aq) + 2e^- \rightarrow H_2(g)$$

In this case, you will notice that one species gains electrons while the other species releases electrons. This is generally the case for oxidation-reduction reactions.

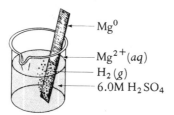

Mg^0

$Mg^{2+}(aq)$
$H_2(g)$
$6.0M\ H_2SO_4$

Equation: $Mg^0(s) + H_2SO_4(aq) = Mg^{2+}(aq) + SO_4^{2-}(aq) + H_2(g)$

Figure 14.3. The Reaction of Magnesium and Sulfuric Acid.

14.7 Cell Potentials

By this time, you might get the idea that we must perform an experiment in order to determine which species will be oxidized and which will be reduced. Fortunately, there is an easier way to determine redox reactions. As you may recall, a voltmeter was inserted in the electrochemical cell in example one. When all connections were made, the voltmeter read a certain potential of the cell. This potential difference between the electrodes in the cell is measured in volts.

By measuring the potential of many cells, a series of values for various electrochemical cells could be obtained. Of course, like any measurement, a reference point is necessary. The reference point for the half-cell reactions of electrochemical cells is the reaction:

$$H_2(g) = 2H^+(aq) + 2e^- \qquad E^0 = 0.00 \text{ volts} \qquad \begin{pmatrix} 1.0 \text{ atm.} \\ 25^\circ \text{ C} \\ 1\ M \end{pmatrix}$$

Thus, hydrogen is used for the standard reference electrode and its standard potential, E^0, is taken as 0.00 volts.

All other half-cell reactions are measured with respect to this half-cell reaction. For example, the reaction between magnesium and 0.5 molar sulfuric acid would be measured to be -2.37 volts. That is, the potential difference between the magnesium electrode and the standard hydrogen electrode is -2.37 volts.

A list of some of the common standard electrode potentials is found in Table 14.1. A more complete list is found in the appendix.

TABLE 14.1
STANDARD ELECTRODE POTENTIALS @ 25°C

Half-Cell	Electrode Potential (E°)
Li^+/Li	−3.04
Na^+/Na	−2.71
Mg^{2+}/Mg	−2.37
Al^{3+}/Al	−1.66
Mn^{2+}/Mn	−1.18
Zn^{2+}/Zn	−0.76
Cr^{3+}/Cr	−0.74
Fe^{2+}/Fe	−0.44
Cd^{2+}/Cd	−0.40
Co^{2+}/Co	−0.28
Ni^{2+}/Ni	−0.25
Sn^{2+}/Sn	−0.14
Pb^{2+}/Pb	−0.13
$2H^+/H_2$	0.00 (definition)
Cu^{2+}/Cu	0.34
Ag^+/Ag	0.80
$Cl_2/2Cl^-$	1.36
$F_2/2F^-$	2.65

A great deal of information may be obtained from Table 14.1. The most important feature is that you can predict which species will be oxidized and which species will be reduced by simply noting their position on the table. This is due to the pattern of the oxidizing and reducing agents.

The strongest oxidizing agents are found in the lower left-hand corner of the chart. That is, the strongest oxidizing agent would be fluorine gas, F_2, followed by Cl_2, and so on up the table.

Fluorine gas, F_2, then will oxidize all other species on the table. The weakest oxidizing agent would be the Li^+ ion followed by Na^+, Mg^{2+}, and so on. That is, the weakest oxidizing agents will be found at the upper left-hand corner of Table 14.1.

Conversely, the strongest reducing agent is Li, followed by Na, Mg, Al, and so on down the chart. The weakest reducing agents would be found in the lower right-hand corner of the table. This means that lithium metal will reduce all chemical species in the table. On the other hand, the fluoride ion, F^-, is not capable of reducing any chemical species.

Another factor on the table is the ability to calculate the potential of the electrochemical cell. This is accomplished by algebraically adding the standard potentials of the half-cell equations. Let us now see how this functions with some examples.

Example 14.5: Determine which species is oxidized, which is reduced, and the potential of a cell composed of 1 M $Zn/Zn^{2+} \| Pb^{2+}/Pb$. Also write the half-cell equations.

Solution:

Step 1: Study the chemical species which compose the cell. Determine which is the stronger oxidizing agent from their position on Table 14.1. The Pb metal is the stronger oxidizing agent. Thus, the lead will be reduced and the zinc will be oxidized. *Remember* the *oxidizing agent is always reduced.* The half-cell equations would be:

$$E^0$$

(oxidized) $Zn^0(s) \rightarrow Zn^{2+}(aq) + 2e^- + 0.76 \text{ volts}$

(reduced) $Pb^{2+}(aq) + 2e^- \rightarrow Pb(s) - 0.13 \text{ volts}$

Step 2: Notice that when the half-cell equation is written in the reverse manner to that found in Table 14.1, the sign of the E^0 value is also reversed, so as to list the proper potential for the equation as written.

Step 3: To obtain the E^0 value for the cell, simply algebraically add the E^0 values of the half-cell equations.

$$E^0_{(oxidized)} = +0.76 \text{ volts}$$
$$E^0_{(reduced)} = -0.13 \text{ volts}$$
$$E^0_{(total)} = +0.63 \text{ volts}$$

Example 14.6: Calculate the E^0 value for a 1 M $Mn/Mn^{2+} \| Cu^{2+}/Cu$ electrochemical cell.

Solution:

Step 1: Determine the species oxidized and the species reduced and write the half-cell equations with proper E^0 values. The Cu^0 metal is the stronger oxidizing agent and therefore, is reduced. The Mn^{2+} is the stronger reducing agent and is therefore, oxidized.

$$E^0$$

(oxidized) $Mn^0(s) = Mn^{2+}(aq) + 2e^- + 1.18 \text{ volts}$

(reduced) $Cu^{2+}(aq) + 2e^- = Cu^0(s) + 0.34 \text{ volts}$

Step 2: Determine the E^0 value by adding the E^0 values of the half-cell equations.

$$E^0_{(oxidized)} \rightarrow +1.18 \text{ volts}$$
$$E^0_{(reduced)} \rightarrow +0.34 \text{ volts}$$
$$E^0_{(total)} \rightarrow +1.52 \text{ volts}$$

As is the case with many topics in chemistry, you will master the operation of balancing redox equations only by a great deal of practice. Whatever you do, however, do not lose sight of the concepts of oxidation and reduction. It is fine to be able to balance equations, but it is the understanding of electrochemical cells and electron transfer which is equally important. It should also be noted that there are many methods for balancing redox equations. Only one method has been presented in this chapter. As a greater knowledge of oxidation-reduction is acquired, innovations and shorter procedures will become apparent.

Glossary

Anode—An electrode in an electrochemical cell at which oxidation occurs.

Cathode—An electrode in an electrochemical cell at which reduction occurs.

Electrochemical Cell—A chemical system in which there is at least one oxidation reaction and one reduction reaction resulting in a cell potential which may be measured in volts.

Electrode—One of the terminals of an electrochemical cell. It is at this location that electrons enter or leave the cell.

Half-Cell Equation—An equation which represents either the oxidation reaction or the reduction reaction of the electrochemical cell.

Oxidation—A chemical process by which a substance undergoes an increase in oxidation state.

Oxidizing Agent—The chemical species which causes oxidation to occur. The oxidizing agent is always reduced.

Redox—An abbreviation for the term oxidation-reduction.

Reduction—A chemical process by which a substance undergoes a decrease in oxidation state.

Reducing Agent—The chemical species which causes reduction to occur. The reducing agent is always oxidized.

Standard Electrode Potential—A measure of the tendency of a specified reduction half-cell reaction to occur. The reference half-cell potential is that of H_2 gas. The potential is expressed in terms of volts for a $1 M$ solution at 1 atm. and $25°C$.

Exercises

1. If a neutral atom becomes negatively charged, has it been oxidized or reduced? Write a general equation using oxygen for the neutral atom.

2. Which of the following is the strongest oxidizing agent? I_2, Ag^+, Cl_2, or F_2. Why?

3. Rank the following in order of decreasing reducing strength. That is, list the strongest reducing agent first and so on. F^-, H_2, Cr, Al, Li.

4. Give an oxidizing agent capable of transforming Mn, to Mn^{2+}.

5. Select a reducing agent capable of transforming Cl_2 to Cl^-.

6. Calculate the E^0 value for the following electrochemical cells:

   a. $Li/Li^+ \| Pb^{2+}/Pb$ c. $Fe^0/Fe^{2+} \| Cu^{2+}/Cu$

   b. $Zn/Zn^{2+} \| Al^{3+}/Al$ d. $Ag/Ag^+ \| Mn^{2+}/Mn$

7. By use of an electrochemical cell, illustrate and explain how you could plate a copper strip with silver. Be sure to label all parts of your diagram.

8. Would it be possible to store a 2 M solution of $Fe(NO_3)_3$ in a container made of lead? Explain your answer in terms of half-cell reactions. Assume the $Fe^{3+}(aq)$ is the oxidizing agent.

9. (a) What would happen if a strip of magnesium were dipped into a solution of $FeSO_4$? (b) What change would occur if a strip of lead were dipped into a solution of $Zn(NO_3)_2$?

10. Balance the following oxidation-reduction equations.

    a. $Cu + NO_3^- = Cu^{2+} + NO_2$ (acid)
    b. $Cu + NO_3^- = Cu^{2+} + NO$ (acid)
    c. $Zn + NO_3^- = Zn^{2+} + N_2$ (acid)
    d. $Zn + NO_3^- = Zn^{2+} + N_2O$ (acid)
    e. $MnO_4^- + H_2S = Mn^{2+} + S$ (acid)
    f. $MnO_4^- + S^{2-} = MnO_2 + S$ (base)
    g. $HOCl + Zn = Cl^- + Zn^{2+}$ (acid)
    h. $H_2SO_4 + HBr = SO_2 + Br_2 + H_2O$
    i. $H_2SO_4 + HI = H_2S + I_2 + H_2O$
    j. $NO_3^- + I_2 + H^+ = IO_3^- + NO_2 + H_2O$
    k. $CuS + NO_3^- = Cu^{2+} + S + NO$ (acid)
    l. $H_2O_2 + MnO_4^- + H^+ = Mn^{2+} + O_2 + H_2O$
    m. $NO_3^- + Zn + OH^- + H_2O = NH_3 + Zn(OH)_4^{2-}$
    n. $ZnS + O_2 = ZnO + SO_2$
    o. $Al + H^+ = Al^{3+} + H_2$

p. $Br_2 + (CO_3)^{2-} = Br^- + BrO_3^- + CO_2$

q. $HBrO = Br^- + O_2$ (acid)

r. $ClO_3^- + I_2 = IO_3^- + Cl^-$ (acid)

s. $NH_3 + O_2 = NO + H_2O$

t. $C + HNO_3 = NO_2 + CO_2 + H_2O$

CHAPTER **15**

Organic Chemistry—I

15.1 Introduction to Organic Chemistry

If a person were to define organic chemistry, he might say it is the study of chemistry relating to substances which contain carbon. Another person might define organic chemistry as the chemistry of living or once living organisms. Either of these definitions would be historically correct, but the real breakthrough in organic chemistry came when Wöhler synthesized the organic substance, urea, from inorganic sources. Thus, the interrelationship of inorganic and organic chemistry was established.

The method of study will be similar to that employed in the inorganic chapters in that the process of naming and formula writing will precede any work involving calculations or reactions. Consequently, Chapter 15 is primarily concerned with naming and formula writing while Chapter 16 deals with reactions, calculations, and bonding. You must be capable of writing formulas and naming organic compounds before you will be able to do anything else in organic chemistry. *Master your nomenclature and formula writing!*

15.2 Alkanes

The simplest class of organic compounds are those compounds which contain only the elements carbon and hydrogen bonded together with single bonds in an open chain structure. This class of compounds is called the *alkanes*. All members of this class of organic compounds will have a name ending in -ane. We shall use the

I.U.P.A.C. Stock System throughout this section. However, common usage will require us to recognize more than one name of a compound at times. Table 15.1 lists some of the common alkanes along with their respective structure and formula.

TABLE 15.1
NOMENCLATURE OF ALKANES

	Structure	Formula	Name
1.	$-\overset{\mid}{\underset{\mid}{C}}-$	CH_4	methane
2.	$-\overset{\mid}{\underset{\mid}{C}}-\overset{\mid}{\underset{\mid}{C}}-$	C_2H_6	ethane
3.	$-\overset{\mid}{\underset{\mid}{C}}-\overset{\mid}{\underset{\mid}{C}}-\overset{\mid}{\underset{\mid}{C}}-$	C_3H_8	propane
4.	$-\overset{\mid}{\underset{\mid}{C}}-\overset{\mid}{\underset{\mid}{C}}-\overset{\mid}{\underset{\mid}{C}}-\overset{\mid}{\underset{\mid}{C}}-$	C_4H_{10}	butane
5.	$-C-C-C-C-C-$	C_5H_{12}	pentane
6.	$-C-C-C-C-C-C-$	C_6H_{14}	hexane
7.	$-C-C-C-C-C-C-C-$	C_7H_{16}	heptane
8.	$-C-C-C-C-C-C-C-C-$	C_8H_{18}	octane
9.	$-C-C-C-C-C-C-C-C-C-$	C_9H_{20}	nonane
10.	$-C-C-C-C-C-C-C-C-C-C-$	$C_{10}H_{22}$	decane

Note: H atoms are not illustrated but are located at the end of each nonterminating bond.

The general formula for any alkane can be derived from the relationship C_nH_{2n+2}, where n is any integer. Using this general formula, the specific formula for any member of the alkane family may be written. For example, the formula of an alkane containing 16 carbon atoms would be $C_{16}H_{2(16)+2} = C_{16}H_{34}$.

15.3 Alkenes

The second class of compounds to consider are those open chained compounds which contain only carbon and hydrogen but have a double bond somewhere in the carbon-carbon chain. When a compound fits this description, it is called an *alkene*. All members of this class of compounds have a name ending in -ene. The -ene ending thus indicates a double bond in a compound containing carbon and hydrogen. Table 15.2 lists some of the common members of the alkene family. The number before the name in some of the examples is used to indicate the bond position of the double bond.

TABLE 15.2
NOMENCLATURE OF ALKENES

Structure	Formula	Name
$>C = C<$	C_2H_4	ethene
$>C = C - C -$	C_3H_6	propene
$>C = C - C - C -$	C_4H_8	1-butene
$>C = C - C - C - C -$	C_5H_{10}	1-pentene
$- C - C = C - C - C -$	C_5H_{10}	2-pentene
$>C = C - C - C - C - C -$	C_6H_{12}	1-hexene
$- C - C = C - C - C - C -$	C_6H_{12}	2-hexene
$- C - C - C = C - C - C -$	C_6H_{12}	3-hexene

The formula for any alkene can be derived from the relationship $C_n H_{2n}$ where n is any integer. This formula does not indicate the position of the double bond, however.

15.4 Alkynes

Another group of hydrocarbons may be classed as alkynes. *Alkynes* are those open chained organic compounds which contain a triple carbon-carbon bond somewhere in the molecule. The -yne ending is characteristic of all members of this class of compounds. Table 15.3 lists some of the common alkynes with their skeleton formula and name.

Notice that the coefficient preceding the name may be the same even though the structure seems different. Actually, the structure is the same in both examples of 2-pentyne. If you could view the formulas from the back side of this page, they would be exactly the

TABLE 15.3
NOMENCLATURE OF ALKYNES

Structure	Formula	Name
$-C \equiv C-$	C_2H_2	ethyne
$-C - C \equiv C-$	C_3H_4	propyne
$-C \equiv C - C - C-$	C_4H_6	1-butyne
$-C - C \equiv C - C-$	C_4H_6	2-butyne
$-C \equiv C - C - C - C-$	C_5H_8	1-pentyne
$-C - C \equiv C - C - C-$	C_5H_8	2-pentyne
$-C - C - C \equiv C - C-$	C_5H_8	2-pentyne

same. In general, always number your bond position from the end which will give the lowest coefficient to the bond position. Thus, you would never have a name such as 3-pentyne. Rather, you would count from the opposite end of the molecule and arrive at 2-pentyne.

15.5 Alcohols

If a hydrogen atom were removed from a member of the alkane series and a hydroxyl radical attached in its place, a class of compounds called *alcohols* would be obtained. All alcohols contain the functional group −OH. This functional group may be attached to any carbon atom in the molecule. Do not think that it is found only on the end of the molecule. In Chapter 16, the preparation of alcohols will be discussed with respect to their production from a given alkane.

Table 15.4 summarizes some of the common alcohols with expanded formulas and correct I.U.P.A.C. names.

Notice that all alcohols have an -ol ending. In the case of the alcohols, a functional group, −OH, is attached to a given compound. The rule of naming is that the functional group must have the lowest coefficient possible. You should now realize that a coefficient before the name of an organic compound can indicate either a bond position or the location of an attached group. This will be illustrated in greater detail later in this chapter.

TABLE 15.4
NOMENCLATURE OF ALCOHOLS

Structure	Formula	Name
$-\overset{\displaystyle |}{C}-OH$	CH_3OH	methanol
$-\overset{\displaystyle |}{C}-\overset{\displaystyle |}{C}-OH$	C_2H_5OH	ethanol
$-\overset{\displaystyle |}{C}-\overset{\displaystyle |}{C}-\overset{\displaystyle |}{C}-OH$	C_3H_7OH	1-propanol
$-\overset{\displaystyle |}{C}-\overset{\displaystyle |}{\underset{\displaystyle OH}{C}}-\overset{\displaystyle |}{C}-$	C_3H_7OH	2-propanol
$-\overset{\displaystyle |}{C}-\overset{\displaystyle |}{C}-\overset{\displaystyle |}{C}-\overset{\displaystyle |}{C}-OH$	C_4H_9OH	1-butanol
$-\overset{\displaystyle |}{C}-\overset{\displaystyle |}{C}-\overset{\displaystyle |}{\underset{\displaystyle OH}{C}}-\overset{\displaystyle |}{C}-$	C_4H_9OH	2-butanol
$-\overset{\displaystyle |}{C}-\overset{\displaystyle |}{C}-\overset{\displaystyle |}{C}-\overset{\displaystyle |}{C}-\overset{\displaystyle |}{C}-OH$	$C_5H_{11}OH$	1-pentanol

15.6 Aldehydes and Ketones

There are two classes of organic compounds which involve a double-bonded oxygen atom. When the oxygen atom and also a hydrogen atom is bonded on the end of the molecule, the compound formed is an *aldehyde*. When the oxygen bonds to a carbon atom somewhere in the middle of the compound, then the class of compounds formed is called a *ketone.*

For convenience, let us study the structure and naming of aldehydes before going into ketones. Table 15.5 lists some of the common aldehydes with their formula and structure.

All aldehydes have the -al ending. Notice that the formulas are written in such a manner so as to emphasize the double bonded oxygen on the end of the molecule. At times this may not be the case. That is, in place of writing C_2H_5CHO you may see the same formula written C_3H_6O and thus you will need additional information before writing the expanded structure or naming the substance because both aldehydes and ketones have the same molecular formula. This is illustrated in Table 15.6.

The ketone family all end in -one. Notice the formula is written in such a manner to emphasize the position of the double-bonded oxygen in the middle of the compound.

Consider the ketone propanone and the aldehyde propanal. Each compound has the formula C_3H_6O. It might be written CH_3COCH_3

TABLE 15.5
NOMENCLATURE OF ALDEHYDES

Structure	Formula	Name
$-C\overset{\displaystyle O}{\underset{\displaystyle H}{\diagdown}}$	CH_2O	methanal
$-C-C\overset{\displaystyle O}{\diagdown H}$	CH_3CHO	ethanal
$-C-C-C\overset{\displaystyle O}{\diagdown H}$	C_2H_5CHO	propanal
$-C-C-C-C\overset{\displaystyle O}{\diagdown H}$	C_3H_7CHO	butanal
$-C-C-C-C-C\overset{\displaystyle O}{\diagdown H}$	C_4H_9CHO	pentanal
$-C-C-C-C-C-C\overset{\displaystyle O}{\diagdown H}$	$C_5H_{11}CHO$	hexanal

TABLE 15.6
NOMENCLATURE OF KETONES

Structure	Formula	Name
$-C-\overset{O}{\overset{\|}{C}}-C-$	CH_3COCH_3	propanone
$-C-C-\overset{O}{\overset{\|}{C}}-C-$	$C_2H_5COCH_3$	2-butanone
$-C-\overset{O}{\overset{\|}{C}}-C-C-$	$CH_3COC_2H_5$	2-butanone
$-C-C-C-\overset{O}{\overset{\|}{C}}-C-$	$C_3H_7COCH_3$	2-pentanone
$-C-C-\overset{O}{\overset{\|}{C}}-C-C-$	$C_2H_5COC_2H_5$	3-pentanone
$-C-C-C-C-\overset{O}{\overset{\|}{C}}-C-$	$C_4H_9COCH_3$	2-hexanone
$-C-C-\overset{O}{\overset{\|}{C}}-C-C-C-$	$C_2H_5COC_3H_7$	3-hexanone

as a ketone and C_2H_5CHO as an aldehyde. This is why it is convenient to write the slightly expanded formula to indicate which compound is being considered.

15.7 Carboxylic Acids

When the functional group –COOH is attached to an alkane, a class of compounds called *carboxylic acids* is produced. In naming carboxylic acids, end the root word with -oic acid. It must be noted that the oxygen and the hydroxide are on the same carbon atom. This carbon atom must be on the end of the molecule. Table 15.7 illustrates some of the common carboxylic acids.

TABLE 15.7
NOMENCLATURE OF CARBOXYLIC ACIDS

Structure	Formula	Name
$-C{\overset{\displaystyle O}{\diagdown}}_{OH}$	HCOOH	methanoic acid
$C-C{\overset{\displaystyle O}{\diagdown}}_{OH}$	CH_3COOH	ethanoic acid (acetic acid)
$C-C-C{\overset{\displaystyle O}{\diagdown}}_{OH}$	C_2H_5COOH	propanoic acid
$C-C-C-C{\overset{\displaystyle O}{\diagdown}}_{OH}$	C_3H_7COOH	butanoic acid
$C-C-C-C-C{\overset{\displaystyle O}{\diagdown}}_{OH}$	C_4H_9COOH	pentanoic acid
$C-C-C-C-C-C{\overset{\displaystyle O}{\diagdown}}_{OH}$	$C_5H_{11}COOH$	hexanoic acid

15.8 Cyclic Compounds

To this point in the study of organic chemistry, we have been concerned with straight-chain compounds. We are now going to consider those compounds which form a ring structure.

The most important, cyclic compound on which our attention

will be focused, will be benzene. Benzene has the formula C_6H_6 and has a structure which has been discussed for a number of years with the result being that there are quite a few ideas on how the bonding occurs. The most generally accepted structure today is that based on molecular orbital theory. In this structure, each carbon-carbon bond is a sigma bond and in addition, there is a pi bond circling the entire ring. Figure 15.1 illustrates this structure.

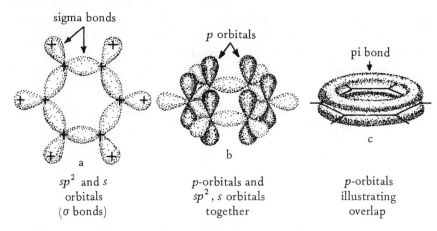

Figure 15.1. Molecular Orbital Structure of Benzene.

For convenience, the structure of benzene is drawn the following way for simplified diagrams.

C_6H_6 (benzene)   bond angle $= 120°$
bond length $= 1.39Å$

There are other cyclic structures of interest and many of them are based on the benzene ring. Table 15.8 lists the structure, formula, and name of some of these compounds.

15.9 Nomenclature of Ethers

All ethers are characterized by an oxygen atom somewhere in the carbon chain. Ethers may have three general formulas. The first general formula is R-O-R′ where the R refers to a given functional group. The second general formula is Ar-O-R where the Ar refers to an aromatic group. The third general formula is Ar-O-Ar which involves two aromatic groups.

TABLE 15.8
NOMENCLATURE OF CYCLIC COMPOUNDS

Structure	Formula	Name
	C_6H_{12}	cyclohexane
	C_6H_6	benzene
	C_6H_{10}	cyclohexene
	$C_{10}H_8$	naphthalene
	$C_{14}H_{10}$	anthracene
	C_5H_{10}	cyclopentane
	C_3H_6	cyclopropane

Ethers are commonly named in one of two ways. The first method is the I.U.P.A.C. system which states that ethers should be named as an alkoxy derivative. The second method is a derived system of nomenclature which names the functional groups and ends with the word ether. As has been the case throughout this chapter, the I.U.P.A.C. system shall be the preferred system. Table 15.9 illustrates the nomenclature and structure of ethers.

TABLE 15.9
NOMENCLATURE OF ETHERS

Structure	Formula	Name
	CH_3OCH_3	methoxymethane (methyl ether)
	$CH_3OC_2H_5$	methoxyethane (methyl-ethyl ether)
	$C_2H_5OC_2H_5$	ethoxyethane (ethyl ether)
	$CH_3OC_3H_7$	methoxypropane (methyl-propyl ether)
	$C_3H_7OC_3H_7$	2-propoxy-2-propane (isopropyl ether)
	$C_6H_5OC_3H_7$	phenoxypropane (phenyl-propyl ether)

15.10 Nomenclature of Esters

The functional group for esters is $R-\overset{\overset{\text{O}}{\|}}{C}-O-R'$. The functional group is not found on the end of a carbon chain.

In naming esters, two rules must be followed. First, determine which carbon chain is in the acid portion of the molecule. Second, name the attached carbon chain in the alcohol portion as a functional group. Remember that all esters end in -oate. Table 15.10 illustrates the nomenclature and structure of esters.

TABLE 15.10
NOMENCLATURE OF ESTERS

Structure	Formula	Name
$-\overset{\|}{\underset{\|}{C}}-\overset{\overset{\text{O}}{\|}}{C}-O-\overset{\|}{\underset{\|}{C}}-$	$CH_3CO_2CH_3$	methylethanoate
$-\overset{\|}{\underset{\|}{C}}-\overset{\overset{\text{O}}{\|}}{C}-O-\overset{\|}{\underset{\|}{C}}-\overset{\|}{\underset{\|}{C}}-$	$CH_3CO_2C_2H_5$	ethylethanoate
$-\overset{\|}{\underset{\|}{C}}-\overset{\|}{\underset{\|}{C}}-O-\overset{\overset{\text{O}}{\|}}{C}-\overset{\|}{\underset{\|}{C}}-\overset{\|}{\underset{\|}{C}}-$	$C_2H_5CO_2C_2H_5$	ethylpropanoate
$-\overset{\|}{\underset{\|}{C}}-\overset{\|}{\underset{\|}{C}}-\overset{\|}{\underset{\|}{C}}-O-\overset{\overset{\text{O}}{\|}}{C}-\overset{\|}{\underset{\|}{C}}-$	$C_3H_7CO_2CH_3$	propylethanoate
$-\overset{\|}{\underset{\|}{C}}-\overset{\|}{\underset{\|}{C}}-\overset{\|}{\underset{\|}{C}}-\overset{\|}{\underset{\|}{C}}-O-\overset{\overset{\text{O}}{\|}}{C}-\overset{\|}{\underset{\|}{C}}-\overset{\|}{\underset{\|}{C}}-$	$C_4H_9CO_2C_2H_5$	butylpropanoate

15.11 The Nomenclature of Functional Groups

For convenience of study, a summary of all the functional classes is made in Table 15.11. Study these as they are the basis for all future nomenclature.

15.12 Nomenclature of Attached Groups

Whenever a group is attached to a straight chain compound or to a cyclic compound, a special name is given to the group so as to be easily identified in a formula or name. Table 15.12 lists the name and formula of some of the common attached groups.

TABLE 15.11
NOMENCLATURE OF FUNCTIONAL GROUPS

Functional Class	Word Ending
alkane	-ane
alkene	-ene
diene	-adiene
alkyne	-yne
alcohol	-ol
aldehyde	-al
ketone	-one
carboxylic acid	-oic acid
ester	-oate
amide	-amide

TABLE 15.12
NOMENCLATURE OF ATTACHED GROUPS

Formula	Name
$-CH_3$	methyl
$-CH_3CH_2$	ethyl
$-CH_3CH_2CH_2$	propyl
$-CH_3CH_2CH_2CH_2$	butyl
$-C_6H_5$ or	phenyl
$-F$	fluoro
$-Cl$	chloro
$-Br$	bromo
$-I$	iodo
$-NO_2$	nitro
$-CN$	cyano (nitrilo)
$-OH$	hydroxyl (when not a functional group)
$-NH_2$	amino

15.13 Examples of I.U.P.A.C. Nomenclature

The following examples are designed to illustrate the various problems which you will encounter in naming organic compounds. Study the structure and then follow the listed rules in naming the compound. The rules for naming organic compounds under the I.U.P.A.C. system are as follows:

Rule 1: Start the root word by selecting the longest continuous chain of carbon atoms which contains the functional group.

Rule 2: Make sure the root word has the ending associated with that particular functional group.

Rule 3: Number the carbons in the chain starting from the end that will allow the carbon atom with the functional group to have as low a number as possible.

Rule 4: Indicate the location of attached groups and special bonds by a coefficient preceeding the name of the group.

Rule 5: If more than one group of the same type is attached, indicate their presence by the prefixes di, tri, tetra, etc., between the coefficient and the name of the attached group. Also, give a position number for each group.

Example 15.1: Name the following organic compounds:

Solution

(a) $-\overset{|}{\underset{|}{C}}-\overset{|}{\underset{|}{C}}-\overset{|}{\underset{|}{C}}-$ (a) propane

(b) $-\overset{|}{\underset{|}{C}}-\overset{|}{\underset{|}{C}}-\overset{|}{\underset{|}{C}}-\overset{|}{\underset{|}{C}}-OH$ (b) 1-butanol

(c) $-\overset{|}{\underset{|}{C}}-\overset{\overset{O}{\overset{||}{}}}{\underset{|}{C}}-\overset{|}{\underset{|}{C}}-$ (c) propanone

(d) $-\overset{|}{\underset{|}{C}}-\overset{|}{\underset{|}{C}}-\overset{|}{\underset{|}{C}}-\overset{|}{\underset{|}{C}}\overset{\nearrow O}{\underset{\searrow H}{}}$ (d) butanoic acid

(e) $-\overset{|}{\underset{|}{C}}-\overset{|}{\underset{|}{C}}-\overset{|}{\underset{|}{C}}-\overset{|}{\underset{|}{C}}-\overset{|}{\underset{|}{C}}\overset{\nearrow O}{\underset{\searrow H}{}}$ (e) pentanol

Example 15.2: Name the following organic compounds:

Solution

(a) $\overset{\diagdown}{\diagup}C=\overset{|}{C}-C\overset{\diagup}{\diagdown}$ (a) propene

(b) $-\overset{|}{\underset{|}{C}}-C\equiv C-\overset{|}{\underset{|}{C}}-$ (b) 2-butyne

(c) $-\overset{|}{\underset{|}{C}}-\overset{|}{\underset{|}{C}}-\overset{|}{\underset{Br}{C}}-\overset{|}{\underset{|}{C}}-\overset{|}{\underset{|}{C}}-$ (c) 3-bromopentane

(d) $-\overset{\underset{|}{\overset{\overset{F}{|}}{C}}}{}-\overset{\underset{\overset{Cl}{|}}{|}}{C}-\overset{|}{\underset{|}{C}}-$

(d) 1-fluoro-2-chloropropane

(e) $-\overset{|}{\underset{|}{C}}-\overset{\underset{\overset{Cl}{|}}{|}}{C}-\overset{\underset{\overset{Cl}{|}}{|}}{C}-\overset{|}{\underset{|}{C}}-OH$

(e) 2, 3-dichloro-1-butanol

Example 15.3: Write formulas for the following compounds:

Solution

(a) ethanol

(a) $-\overset{|}{\underset{|}{C}}-\overset{|}{\underset{|}{C}}-OH$

(b) benzene

(b)

(c) phenol

(c) $-OH$

(d) propanol

(d) $-\overset{|}{\underset{|}{C}}-\overset{|}{\underset{|}{C}}-C\overset{\displaystyle O}{\underset{\displaystyle H}{\big\langle}}$

(e) 2-hexanone

(e) $-\overset{|}{\underset{|}{C}}-\overset{\overset{\displaystyle O}{\|}}{C}-\overset{|}{\underset{|}{C}}-\overset{|}{\underset{|}{C}}-\overset{|}{\underset{|}{C}}-\overset{|}{\underset{|}{C}}-$

or

$-\overset{|}{\underset{|}{C}}-\overset{|}{\underset{|}{C}}-\overset{|}{\underset{|}{C}}-\overset{|}{\underset{|}{C}}-\overset{\underset{\displaystyle O}{\|}}{C}-\overset{|}{\underset{|}{C}}-$

Example 15.4: Write structures for the following compounds:

(a) 2-methylpropane

(a) $-\overset{|}{\underset{\underset{\textstyle -C-}{|}}{C}}-C=C\big\langle$

(b) 1, 2-dichloro-3-ethylhexane

(b) $-\overset{|}{\underset{\overset{\displaystyle Cl}{|}}{C}}-\overset{\overset{\displaystyle Cl}{|}}{\underset{|}{C}}-\overset{|}{\underset{\underset{\textstyle -C-}{|}}{C}}-\overset{|}{\underset{|}{C}}-\overset{|}{\underset{|}{C}}-\overset{|}{\underset{|}{C}}-$

(c) 3-bromo-2-butanol

(c) $-\overset{|}{\underset{|}{C}}-\overset{\overset{\displaystyle Br}{|}}{\underset{|}{C}}-\overset{|}{\underset{\overset{\textstyle OH}{}}{C}}-\overset{|}{\underset{|}{C}}-$

(d) 1-chloro-2-butanone

(d) $-\overset{|}{\underset{|}{C}}-\overset{|}{\underset{|}{C}}-\overset{\overset{O}{\|}}{C}-\overset{Cl}{\underset{|}{C}}-$

(e) 3, 3-dimethyl-1-pentene

(e) $-\overset{|}{\underset{|}{C}}-\overset{|}{\underset{|}{C}}-\overset{\overset{-\overset{|}{C}-}{|}}{\underset{\underset{-\overset{|}{C}-}{|}}{C}}-\overset{|}{C}=C\diagdown^{\diagup}$

Example 15.5: Give the correct I.U.P.A.C. name for the following compounds:

compounds:	Solution
(a) $-\overset{|}{\underset{|}{C}}-\overset{\overset{Br}{|}}{\underset{|}{C}}-\overset{|}{\underset{Cl}{C}}-C\overset{\diagup O}{\diagdown OH}$	(a) 2-chloro-3-bromobutanoic acid
(b) $-\overset{\overset{NO_2}{|}}{\underset{|}{C}}-\overset{\overset{C}{|}}{\underset{F}{C}}-C\overset{\diagup O}{\diagdown}$	(b) 2-fluoro-2-methyl-3-nitropropanal
(c) $-\overset{|}{\underset{|}{C}}-\overset{|}{\underset{|}{C}}-\overset{|}{\underset{OH}{C}}-\overset{\overset{Cl}{|}}{\underset{Cl}{C}}-\overset{|}{\underset{|}{C}}-OH$	(c) 2,2-dichloro-1,3-pentadiol

Example 15.6: Write the correct I.U.P.A.C. name for the following cyclic compounds:

cyclic compounds:·   Solution

(a)   (a) phenol

(b)   (b) 1, 3-dinitrobenzene

(c)   (c) 1, 4-dichlorobenzene

(d)

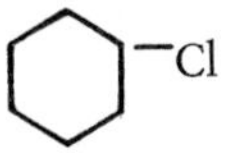

(d) chlorocyclohexane

(e)

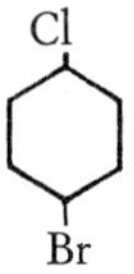

·(e) 1-bromo-4-chlorocyclohexane

(f)

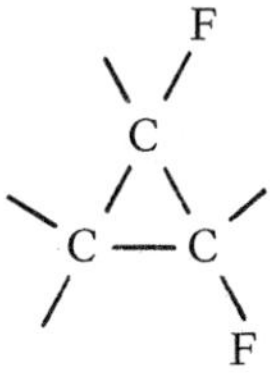

(f) 1, 2-difluorocyclopropane

(g)

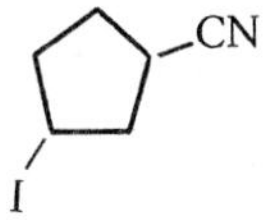

(g) 1-cyano–2-iodocyclopentane

Glossary

Alcohol–A class of organic compounds containing the functional group -OH.

Aldehyde–A class of organic compounds containing a carbonyl group -CHO.

Alkane–A class of open chained organic compounds containing only carbon and hydrogen with only single chemical bonds throughout the molecule.

Alkene–A class of open chained organic compounds containing one carbon-carbon double bond somewhere in the molecule.

Alkyne–A class of open chained organic compounds containing one carbon-carbon triple bond somewhere in the molecule.

Carboxylic Acid–A class of organic compounds containing a $-C\overset{\displaystyle O}{\underset{\displaystyle OH}{\Big\langle}}$ group.

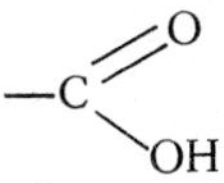

Cyclic Compounds–That group of organic compounds which forms a ring structure. It must contain three or more carbon atoms.

Ketones–A class of organic compounds characterized by a carbonyl group, $-\overset{O}{\overset{\|}{C}}-$, attached to alkyl or aryl groups on both sides.

Exercises

1. Classify each of the following according to their functional groups such as alkane, alkene, alcohol, etc.

   a. 2-propanol
   b. acetic acid
   c. 2-butanone
   d. 2-pentene
   e. ethanal

2. Classify each of the following according to their functional groups.

a. $-\overset{|}{\underset{|}{C}}-\overset{|}{\underset{|}{C}}-\overset{|}{C}=C\diagdown$

b. $-\overset{|}{\underset{|}{C}}-C\equiv C-$

c. $-\overset{|}{\underset{|}{C}}-\overset{\overset{O}{\parallel}}{C}-\overset{|}{\underset{|}{C}}-$

d. $-\overset{|}{\underset{|}{C}}-\overset{|}{\underset{|}{C}}-\overset{|}{\underset{|}{C}}-\overset{|}{\underset{|}{C}}-C\diagup^{O}_{\diagdown OH}$

e. $-\overset{|}{\underset{|}{C}}-\overset{|}{\underset{|}{C}}-\overset{|}{\underset{|}{C}}-OH$

3. Write expanded structural formulas for:

a. $CH_3 CH_2 CH_2 CH_2 CH_3$ c. $CH_3 COCH_2 CH_3$

b. $(CH_3)_2 CHCH_3$ d. $C_6 H_6$

4. Name the following attached groups using I.U.P.A.C. nomenclature.

a. $-F$ d. $-NO_2$

b. $-NH_2$ e. $-OH$

c. $-CN$

5. Write the correct I.U.P.A.C. name for each of the following:

a. $-\overset{|}{C}-\overset{\overset{O}{\parallel}}{C}-\overset{|}{C}-$

b. $-C\equiv C-$

c. $-\overset{|}{\underset{|}{C}}-\overset{|}{\underset{|}{C}}-\overset{|}{\underset{|}{C}}-C\diagup^{O}_{\diagdown}$

d. $-\overset{|}{C}-\overset{|}{C}-\overset{|}{C}-OH$

e. $-\overset{|}{\underset{|}{C}}-\overset{|}{\underset{|}{C}}-\overset{|}{\underset{|}{C}}-C\diagup^{O}_{\diagdown OH}$

6. Write the correct I.U.P.A.C. name for each of the following:

a. $-\overset{\overset{Br}{|}}{C}-\overset{|}{\underset{\underset{Br}{|}}{C}}-\overset{|}{C}-OH$

b. $-\overset{|}{\underset{\underset{F}{|}}{C}}-\overset{|}{C}=\overset{|}{C}-C\diagup^{F}_{\diagdown}$

c. $Br-C\equiv C-Br$

d. $-\overset{|}{C}-\overset{\overset{O}{\parallel}}{C}-\overset{|}{C}-Br$
   $-\overset{|}{\underset{\underset{Cl}{|}}{C}}-\quad-\overset{|}{\underset{|}{C}}-$

e. $-\overset{|}{C}-\overset{\overset{CN}{|}}{\underset{|}{C}}-\overset{|}{C}-\overset{|}{\underset{\underset{NO_2}{|}}{C}}-C\diagup^{O}_{\diagdown}$

7. Write the correct I.U.P.A.C. name for each of the following:

a. NO$_2$

b.

c.

e.   OH

d.   NO$_2$

8. Write correct structural formulas for the following:
   a.   2, 3-dinitro-1-pentene   d.   3, 3, 4, 4-tetrachlorohexane
   b.   1, 3-dinitrobutane   e.   2, 3-dimethylhexane
   c.   1, 4-dichlorobenzene   f.   benzoic acid

9. Write correct structural formulas for the following:
   a.   2-methyl-2-hexene
   b.   2, 3, 4-trimethyl-2-heptene
   c.   2, 3-diphenyl-4-nitrooctane
   d.   2-phenylbutane
   e.   3-iodo-1-propene
   f.   2, 3-dichloropropanal

10. Write correct structural formulas for the following:
   a.   2-bromo-2, 3-diphenylbutane
   b.   2, 3-dinitro-1-propene
   c.   2, 2, 3, 3-tetraiodopentane
   d.   3-chlorophenol
   e.   2, 3-difluorobutanal
   f.   tetrachloromethane

CHAPTER 16

Organic Chemistry—II

16.1 Introduction

In Chapter 15, a brief presentation was given on the various families which compose much of organic chemistry. The key element in all cases, and the building block of organic chemistry, is carbon. In the following sections, bonds involving oxygen, the halogens, and other attached groups will be discussed; but it is the bonding characteristics of carbon that are of major importance.

16.2 Bonding of Carbon

From our previous work on structure and bonding, it should be remembered that carbon occupies a unique slot in electron structure. As a review, carbon has atomic number six and therefore, has six electrons. The electron configuration would be $1s^2 2s^2 2p^2$. Thus, there are four valence electrons in carbon, and in terms of bonding, carbon has the tendency to form four covalent bonds by sharing its four electrons with four from other bonding species. If four other bonds are obtained, carbon has completed its valence electron requirement of eight electrons. The eight electrons occupy four bonding orbitals. The shapes of the orbitals are determined by the atoms involved in the bond.

Since the carbon bond is of such importance in organic chemistry, a great deal of information is known about its bonding parameters; such things as the bond angle, the bond distance, the molecular shape and the bond energy.

The bond angle for various molecules involving carbon vary greatly. Do not think that carbon only forms the tetrahedral shaped molecule and thus always has a bond angle of 109.5°. This is a common bond angle for carbon, but certainly not the only one. Table 16.1 lists the bond angle for some molecules containing carbon.

TABLE 16.1
BOND ANGLES ABOUT THE CARBON ATOM

Molecule	Angle in Degrees
CH_4	109.5
CH_3Cl	110.5 (H–C–Cl)
C_2H_6	109.3 (H–C–H)
CH_3OH	109.3 (H–C–OH)
CO_2	180

The bond length or the average distance between the nuclei of two bonded atoms also varies from molecule to molecule in situations which involve carbon. In our earlier discussion of bonding, it was mentioned that a double and triple bond will tend to be a stronger bond than a single bond. Notice that when carbon forms a double bond, the bond distance is decreased from that of a single bond involving the same species. Also notice the relationship of the triple bond. Table 16.2 lists the bond distance for some molecules involving carbon.

A parameter closely related to the bond angle and the bond distance is the structure or molecular shape. The molecular shape is,

TABLE 16.2
VARIATIONS OF BOND LENGTHS
FOR SOME CARBON BONDS

Bond	Molecule	Bond Length (Å)
C–C	Diamine	1.54
C–C	Ethane	1.54
C–C	Ethanol	1.55
C–C	Benzene	1.39
C–H	Methane	1.10
C–H	Ethane	1.10
C=C	Ethene	1.34
C≡C	Ethyne	1.20
C–H	Ethyne	1.09

of course, dependent on the bonding orbitals of the species involved. From Table 16.1 and Table 16.2, it is evident that there are a few general patterns with respect to the shapes obtained for molecules containing carbon.

If we were to study the types of molecular orbitals formed by carbon and its bonding species, we would observe two types of orbital bonding.

The first type of bonding orbital formed is called a sigma (σ) bonding orbital. In short, a sigma bond occurs when the two electrons forming the covalent bond are located along the axis of the molecule. That is, the two electrons are in the region of space between the two bonded atoms. Figure 16.1 illustrates a sigma bond between a carbon atom and a hydrogen atom.

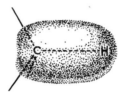

Figure 16.1. Sigma Orbital
Bond in a Hydrocarbon
Molecule.

A classic example of the use of sigma bonds in a hydrocarbon molecule is methane, CH_4. All four bonds in the methane molecule are sigma bonds originating when sp^3 atomic orbitals of carbon bond with s atomic orbitals of hydrogen to form sigma molecular orbitals for the methane molecule. Figure 16.2 illustrates this bonding graphically.

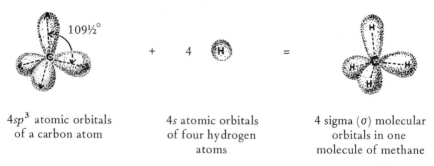

$4sp^3$ atomic orbitals of a carbon atom	$4s$ atomic orbitals of four hydrogen atoms	4 sigma (σ) molecular orbitals in one molecule of methane

Figure 16.2. Sigma Bonding in the Methane Molecule.

The second type of bonding molecular orbital formed is called a pi (π) bonding orbital. A pi bond occurs when the two electrons forming the covalent bond are located above and below the axis of the molecule. Very often, a p atomic orbital is involved when you have a π bond. This is because the shape of the p orbital allows the electron to be located above and below the axis of the molecule.

If two p_z atomic orbitals were to bond, a pi (π) bonded molecular orbital would occur. This is illustrated in Figure 16.3.

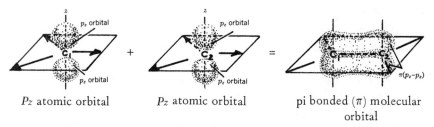

P_z atomic orbital P_z atomic orbital   pi bonded (π) molecular orbital

Figure 16.3. An Example of a Pi-Bonded Molecular Orbital.

Since carbon generally forms the sp^3 hybrid atomic orbital, it is apparent that for carbon to form a pi molecular bond, the other bonding species must arrange itself in such a manner to have maximum overlap in the region above and below the axis of the sigma bond. For this reason, a pi bond will usually involve a p atomic orbital from the noncarbon bonding species.

A few additional examples may help to clarify this relationship and also the relationship between the sigma and the pi bond. Figure 16.4 illustrates examples of sigma bonds, pi bonds, and a combination of sigma and pi bonds in carbon bonded molecules.

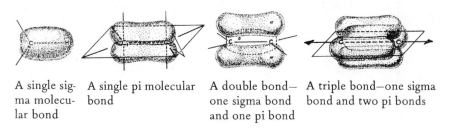

A single sigma molecular bond   A single pi molecular bond   A double bond—one sigma bond and one pi bond   A triple bond—one sigma bond and two pi bonds

Figure 16.4. Illustrations of Sigma and Pi Bonding.

In summary, remember that the most common shape for the molecules involving carbon is the tetrahedral shape. However, there are other shapes possible depending on the bonding species. In terms of molecular bonds, we have two types. First, the sigma (σ) bond in the region between the two bonding atoms and secondly, the pi (π) bond with regions of overlap above and below the two bonding species. Lastly, remember that carbon forms covalent bonds and thus there must always be a sharing of electrons.

16.3 Isomerism

An interesting phenomenon in chemistry is isomerism. *Isomerism* is the existence of two or more chemical species with the same molecular formula and molecular mass but they do not have the same physical properties and quite often they react very differently.

There are different ways of discussing isomerism depending upon which physical properties are of interest. The four most common types of isomerism are positional isomerism, skeletal isomerism, functional isomerism, and optical isomerism. Let us now look at each of these types of isomerism in detail.

Positional isomerism is dependent upon a functional group. In a straight chained carbon compound, a functional group may be attached to any one of a number of carbon atoms. Different positional isomers are obtained when the functional group is attached to various carbon atoms. This can best be illustrated with an example. Consider the molecular formula $C_5 H_{11} OH$. Let us assume we are only considering the family of alcohols and let us write all possible molecular structures for this formula. From this, the concept of positional isomerism can be seen. That is, the OH group which is the functional group will move from carbon atom to carbon atom forming different molecular structures, yet at all times, retaining the same molecular formula. Since we have restricted the formula $C_5 H_{11} OH$ to the alcohol family, Figure 16.5 illustrates the possible positional isomers for a straight carbon chain.

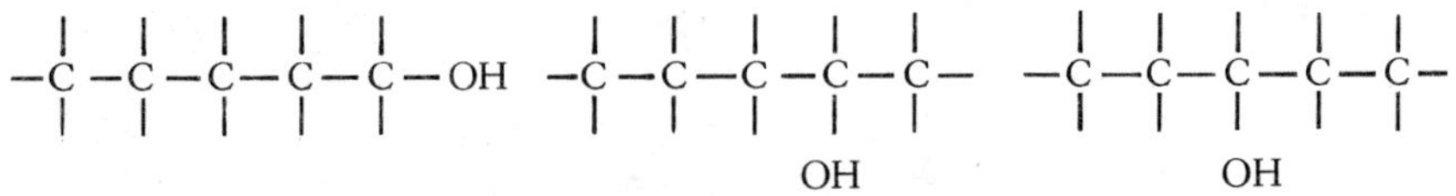

Figure 16.5. Illustration of Positional Isomerism.

Skeletal isomerism is a type of isomerism based on the branching of the carbon chain. In other words, a straight chained compound is branched into a series of shorter chains of carbon atoms. An example of skeletal isomerism is that of the four-carbon alcohols. All isomers have the molecular formula C_4H_9OH but differ greatly in their molecular structure. Figure 16.6 illustrates the various skeletal isomers for the molecular formula C_4H_9OH. Remember we are referring only to the alcohol family with this example.

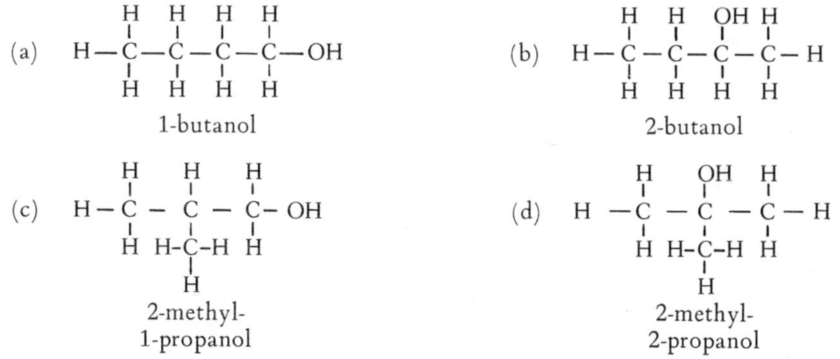

(a)

$$H-\overset{\overset{\displaystyle H}{|}}{\underset{\underset{\displaystyle H}{|}}{C}}-\overset{\overset{\displaystyle H}{|}}{\underset{\underset{\displaystyle H}{|}}{C}}-\overset{\overset{\displaystyle H}{|}}{\underset{\underset{\displaystyle H}{|}}{C}}-\overset{\overset{\displaystyle H}{|}}{\underset{\underset{\displaystyle H}{|}}{C}}-OH$$

1-butanol

(b)

$$H-\overset{\overset{\displaystyle H}{|}}{\underset{\underset{\displaystyle H}{|}}{C}}-\overset{\overset{\displaystyle H}{|}}{\underset{\underset{\displaystyle H}{|}}{C}}-\overset{\overset{\displaystyle OH}{|}}{\underset{\underset{\displaystyle H}{|}}{C}}-\overset{\overset{\displaystyle H}{|}}{\underset{\underset{\displaystyle H}{|}}{C}}-H$$

2-butanol

(c)

$$H-\overset{\overset{\displaystyle H}{|}}{\underset{\underset{\displaystyle H}{|}}{C}}-\overset{\displaystyle H}{\underset{\underset{\displaystyle H-C-H}{|}}{C}}-\overset{\overset{\displaystyle H}{|}}{\underset{\underset{\displaystyle H}{|}}{C}}-OH$$
$$\underset{\displaystyle H}{}$$

2-methyl-
1-propanol

(d)

$$H-\overset{\overset{\displaystyle H}{|}}{\underset{\underset{\displaystyle H}{|}}{C}}-\overset{\overset{\displaystyle OH}{|}}{\underset{\underset{\displaystyle H-C-H}{|}}{C}}-\overset{\overset{\displaystyle H}{|}}{\underset{\underset{\displaystyle H}{|}}{C}}-H$$
$$\underset{\displaystyle H}{}$$

2-methyl-
2-propanol

Figure 16.6. Positional and Skeletal Isomerism for Alcohols with the Formula C_4H_9OH.

In the discussion of the first two types of isomerism, we have limited the discussion to the alcohol family to simplify the discussion. Now, however, let us consider *functional isomerism. Functional isomerism* deals with chemical species which have the same chemical formula but different functional groups. In other words, we are no longer limiting the discussion to a single family but rather considering all the possible families or functional classes. As an example, consider the chemical formula $C_4H_{10}O$. This molecular formula could represent the alcohol family as seen in Figure 16.6, but it might also represent the ether family. The ether family has an oxygen atom bonded between two carbon atoms on the chain. Figure 16.7 illustrates some functional isomers for the chemical formula $C_4H_{10}O$.

The next type of isomerism we shall consider is that of stereo isomerism. There are two types of stereo isomerism. One is called geometric isomerism and the other is called optical isomerism. *Optical isomerism* is dependent on the attached groups of a given compound. There are times, due to the fact that space is three-dimension-

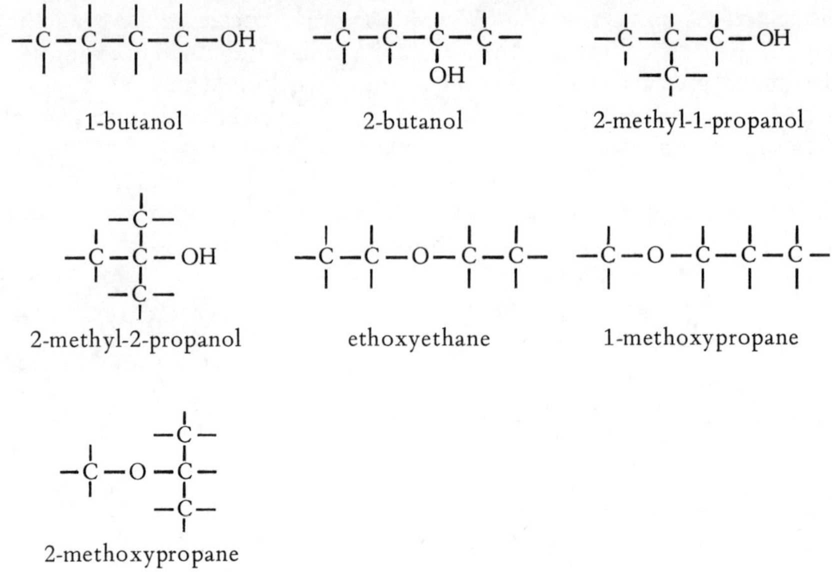

$$-\overset{|}{\underset{|}{C}}-\overset{|}{\underset{|}{C}}-\overset{|}{\underset{|}{C}}-\overset{|}{\underset{|}{C}}-OH$$

1-butanol

$$-\overset{|}{\underset{|}{C}}-\overset{|}{\underset{|}{C}}-\overset{|}{\underset{|}{C}}-\overset{|}{\underset{OH}{C}}-$$

2-butanol

$$-\overset{|}{\underset{|}{C}}-\overset{|}{\underset{\underset{-C-}{|}}{C}}-\overset{|}{\underset{|}{C}}-OH$$

2-methyl-1-propanol

$$-\overset{|}{\underset{|}{C}}-\overset{\overset{-C-}{|}}{\underset{\underset{-C-}{|}}{C}}-OH$$

2-methyl-2-propanol

$$-\overset{|}{\underset{|}{C}}-\overset{|}{\underset{|}{C}}-O-\overset{|}{\underset{|}{C}}-\overset{|}{\underset{|}{C}}-$$

ethoxyethane

$$-\overset{|}{\underset{|}{C}}-O-\overset{|}{\underset{|}{C}}-\overset{|}{\underset{|}{C}}-\overset{|}{\underset{|}{C}}-$$

1-methoxypropane

$$-\overset{|}{\underset{|}{C}}-O-\overset{\overset{-C-}{|}}{\underset{\underset{-C-}{|}}{C}}-$$

2-methoxypropane

Figure 16.7. Positional and Functional Isomers for $C_4H_{10}O$.

al, that a species which has exactly the same functional groups attached to the same carbons is not identical in chemical behavior or in structure. They are, in fact, mirror images of each other. That is, if you were to look at the structure of a molecule in a mirror, you would see the other isomer of that species. Optical isomerism must occur in pairs. You have the original molecule and its mirror image. A convenient example is 2-butanol. Figure 16.8 illustrates the optical isomerism of 2-butanol.

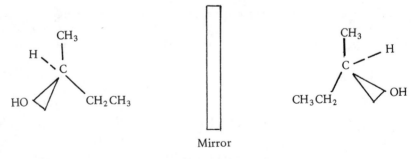

Mirror

Figure 16.8. Optical Isomers of 2-Butanol.

No matter how you translate or rotate the structures, they are not superimposable. They are, however, mirror images. Additional treatment of optical isomerism can be found in Chapter 17.

Isomers will generally have different chemical properties due to the variations in structures. For example, it is much easier to oxidize a primary alcohol such as 1-butanol than it is a secondary alcohol such as 2-methyl-1-propanol. Accordingly, it is easier to oxidize 2-methyl-1-propanol than it is the tertiary alcohol 2-methyl-2-propanol. In fact, the tertiary alcohol does not oxidize at all when reacted with $KMnO_4$. These variations in reactivity will be discussed in the following sections in much greater detail.

16.4 Reactions of the Various Families

In studying the reactions of a particular family in organic chemistry, you will find a pattern established in most cases. No attempt will be made to list all types of reactions for a given family. Rather, a few classical reactions will be considered.

16.5 Reactions of Alcohols and Phenols

Both the alcohols and the phenols have an OH group as their functional group. The distinction between the two families is that an alcohol has an alkyl group attached to the —OH group and the phenol has an aryl group attached to the —OH group. As a convenient shorthand system, the alcohols can generally be represented by the formula R—OH and the phenols by Ar—OH.

The first reaction we shall consider is the reaction of an active metal such as sodium with an alcohol and also with a phenol. In terms of a stepwise reaction, the first step in the reaction of an alcohol with an active metal is the formation of an alkoxide ion from the dissociation of the alcohol. This can be illustrated by the equation:

$$\text{ROH} \overset{\text{dissociation}}{\rightleftharpoons} \text{RO}^- + \text{H}^+$$
$$\text{alkoxide ion}$$

The hydrogen ion in turn, then reacts with the active metal forming a metallic alkoxide. In general, the second step in the reaction is:

$$H^+ + RO^- + M \rightleftharpoons RO^- M^+ + 1/2\,H_2\,(g)$$
(alkoxide)   (metal)   metallic alkoxide

To clarify this reaction, consider a specific example such as the reaction of ethanol and sodium metal. Step one of the formation of the alkoxide ion.

Reaction 1:

$$-\overset{|}{\underset{|}{C}}-\overset{|}{\underset{|}{C}}-OH \rightleftharpoons -\overset{|}{\underset{|}{C}}-\overset{|}{\underset{|}{C}}-O^- + H^+$$
ethanol   ethoxide ion

Reaction 2:

$$H^+ + -\overset{|}{\underset{|}{C}}-\overset{|}{\underset{|}{C}}-O^- + Na \rightleftharpoons -\overset{|}{\underset{|}{C}}-\overset{|}{\underset{|}{C}}-O^-\!\cdot\!-Na^+ + 1/2\,H_2\,(g)$$
ethoxide ion   sodium   sodium ethoxide

Total reaction:

$$2-\overset{|}{\underset{|}{C}}-\overset{|}{\underset{|}{C}}-OH + 2Na \rightleftharpoons 2-\overset{|}{\underset{|}{C}}-\overset{|}{\underset{|}{C}}-ONa + H_2\,(g)$$

Step two was the formation of the metallic alkoxide, in this case, sodium ethoxide.

When a phenol is mixed with an active metal, a reaction occurs. The reason for this in simple terms is that phenols are acidic. Thus, when sodium is combined with phenol, a typical acid–active metal reaction occurs. In equation form the reaction of alcohols and phenols can be summarized as follows:

2ROH	+	2M $\rightleftharpoons$ 2ROM	+	$H_2\,(g)$
alcohol		metal metallic		
		alkoxide		
2ArOH	+	2M $\rightleftharpoons$ 2ArOM	+	$H_2\,(g)$
phenol		metal		

A closely related reaction to the previous case is the reaction of a metallic hydroxide with an alcohol and also with a phenol. As was mentioned in the previous section, phenols are more acidic than alcohols. In terms of pH, alcohols are generally neutral. Thus, the likelihood of an alcohol reacting with a metallic hydroxide which is quite basic, is remote. In fact, when sodium hydroxide is combined with methanol, there is no reaction. In equation form:

$$CH_3OH \quad + \quad NaOH \quad \rightarrow \quad \text{no reaction}$$
$$\text{methanol} \qquad \text{sodium hydroxide}$$

In general, we can state that an alcohol does not react with a metallic base under normal laboratory conditions.

Since the phenols are more acidic than alcohols, they will react with the metallic hydroxides. In general form, the reaction is:

$$ArOH \quad + \quad MOH \quad \rightarrow \quad ArO^-M^+ \quad + \quad H_2O$$
$$\text{phenol} \qquad \text{metallic} \qquad \text{metallic}$$
$$\text{hydroxide} \qquad \text{phenoxide}$$

Notice the bond between the oxygen and the metal in the phenoxide. The oxygen remains slightly electronegative and the metal is electropositive. In future reactions, we shall see how this bond can be broken and new products conveniently formed. As a specific example, consider the reaction of sodium hydroxide and phenol.

$$\text{phenol} \qquad \text{sodium} \qquad \text{sodium}$$
$$\text{hydroxide} \qquad \text{phenoxide}$$

In general, we can state that a phenol reacts with a metallic hydroxide to form a phenoxide and water.

To summarize this information in equation form, we have:

$$ROH \quad + \quad MOH \quad \rightarrow \quad \text{no reaction}$$
$$\text{alcohol} \qquad \text{metallic}$$
$$\text{hydroxide}$$

$$ArOH \quad + \quad MOH \quad \rightarrow \quad ArO^-M^+ + H_2O$$
$$\text{phenol} \qquad \text{metallic} \qquad \text{metallic}$$
$$\text{hydroxide} \qquad \text{phenoxide}$$

Another reaction of interest is the oxidation of alcohols. Two common oxidizing agents used to oxidize alcohols are potassium permanganate ($KMnO_4$) and potassium dichromate ($K_2Cr_2O_7$). Oxidation of the alcohol amounts to the removal of the hydrogen ion from the functional group and the resulting establishment of a double bonded oxygen at that position.

When a *primary* alcohol, or in other words an alcohol which has the functional group attached to an end carbon of the chain, is oxidized, an aldehyde is produced. The general equation would be:

$$R-\underset{\underset{\text{alcohol}}{\text{primary}}}{\overset{|}{\underset{|}{C}}}-OH \quad \underset{K_2Cr_2O_7}{\overset{KMnO_4}{\underset{\text{or}}{\rightarrow}}} \quad R-C\overset{\nearrow O}{\underset{\searrow}{}} \quad +H_2O$$
$$\text{aldehyde}$$

In more specific terms, consider the oxidation of 1-butanol to form butanal.

$$\underset{\text{1-butanol}}{-\overset{|}{\underset{|}{C}}-\overset{|}{\underset{|}{C}}-\overset{|}{\underset{|}{C}}-\overset{|}{\underset{|}{C}}-OH} \quad \underset{K_2Cr_2O_7}{\overset{KMnO_4}{\underset{\text{or}}{\rightarrow}}} \quad \underset{\text{butanal}}{-\overset{|}{\underset{|}{C}}-\overset{|}{\underset{|}{C}}-\overset{|}{\underset{|}{C}}-C=O} + H_2O + \text{etc.}$$

products depending on the oxidizing agents

When a *secondary* alcohol or in other words, an alcohol which has the functional group attached to a nonterminating carbon in the chain is oxidized, the product formed is a ketone. In equation form the reaction for a secondary alcohol such as 2-butanol would be as follows:

$$\underset{\text{2-butanol}}{-\overset{|}{\underset{|}{C}}-\overset{|}{\underset{|}{C}}-\overset{|}{\underset{\underset{OH}{|}}{C}}-\overset{|}{\underset{|}{C}}-} \quad \underset{K_2Cr_2O_7}{\overset{KMnO_4}{\underset{\text{or}}{\rightarrow}}} \quad \underset{\text{butanone}}{-\overset{|}{\underset{|}{C}}-\overset{|}{\underset{|}{C}}-\overset{|}{\underset{\underset{O}{||}}{C}}-\overset{|}{\underset{|}{C}}-} + \text{etc.}$$

Notice that the mechanics of the reaction are the same with the difference in the product, caused by the location of the functional OH group.

Under normal laboratory conditions, the oxidizing agents $KMnO_4$ and $K_2Cr_2O_7$ will not oxidize a tertiary alcohol. Remember that a *tertiary* alcohol has a branched structure and consequently, is resistant to oxidation. In summary, the following word equations can be written for the family of alcohols:

Primary Alcohol	$\underset{K_2Cr_2O_7}{\overset{KMnO_4}{\underset{\text{or}}{\rightarrow}}}$	Aldehyde
Secondary Alcohol	$\underset{K_2Cr_2O_7}{\overset{KMnO_4}{\underset{\text{or}}{\rightarrow}}}$	Ketone
Tertiary Alcohol	$\underset{K_2Cr_2O_7}{\overset{KMnO_4}{\underset{\text{or}}{\rightarrow}}}$	no reaction

16.6 Reactions of Alkenes

The alkene family is of special interest because of the double bond in the molecule. When reactions occur with alkenes, you will notice that the double bond position is the active position. That is, it is at this position that a change in structure will occur. As an initial reaction, consider the reaction of an alkene with an acid of the general formula HX. Examples of such acids would be HF, HCl, HBr, HI, and H_2SO_4. Upon reaction, the double bond is broken and a haloalkane is formed. It is important to notice the position of attack. The halogen will attach to a carbon according to *Markownikoff's Rule*. This rule states—In the addition of an acid, HX, to an unsymmetric alkene, the hydrogen from the acid attaches to that carbon atom of the double bond which already has the greater number of hydrogen atoms attached to it. The X group attaches to the other carbon atom of the double bond. Like any rule, it is the general pattern but there are exceptions so do not think this rule of addition works in all possible situations. To gain further insight on these addition reactions, study the following examples.

(a) $\ce{>C=C<}$ ethene $\xrightarrow{\text{HCl}}$ $-\overset{|}{\underset{|}{C}}-\overset{|}{\underset{|}{C}}-Cl$ chloroethane

Note: point of attachment makes no difference as this is a symetrical compound.

(b) $-\overset{|}{\underset{|}{C}}-\overset{|}{C}=C\backslash$ propene $\xrightarrow{\text{HBr}}$ $-\overset{|}{\underset{|}{C}}-\overset{|}{\underset{Br}{C}}-\overset{|}{\underset{|}{C}}-$ 2-bromopropane

Note: attachment of hydrogen and bromine

(c) $-\overset{|}{\underset{|}{C}}-\overset{|}{\underset{|}{C}}-\overset{|}{C}=C\backslash$ 1-butene $\xrightarrow{\text{HI}}$ $-\overset{|}{\underset{|}{C}}-\overset{|}{\underset{|}{C}}-\overset{|}{\underset{I}{C}}-\overset{|}{\underset{|}{C}}-$ 2-iodobutane

Note: attachment of hydrogen and iodine

(d) $-\overset{|}{\underset{|}{C}}-\overset{|}{C}=\overset{|}{C}-C\backslash$ 2-butene $\xrightarrow{\text{HCl}}$ $-\overset{|}{\underset{|}{C}}-\overset{|}{\underset{Cl}{C}}-\overset{|}{\underset{|}{C}}-\overset{|}{\underset{|}{C}}-$

or

$-\overset{|}{\underset{|}{C}}-\overset{|}{\underset{|}{C}}-\overset{|}{\underset{Cl}{C}}-\overset{|}{\underset{|}{C}}-$ 2-chlorobutane

} same product

(e) $-\overset{|}{\underset{|}{C}}-\overset{|}{C}=C\overset{\diagup}{\diagdown}$ $\xrightarrow{H_2SO_4}$ $-\overset{|}{\underset{|}{C}}-\overset{|}{\underset{|}{C}}-\overset{|}{\underset{|}{C}}-$

propene

$$\overset{\displaystyle O}{\underset{\displaystyle\underset{\displaystyle H}{\underset{\displaystyle |}{O-S-O}}}{|}}$$

2-propanesulfonic acid

16.7 Reactions of Alkanes

The alkanes contain only carbon and hydrogen with single chemical bonds in an open chain. Thus, one might expect the reactions of the alkanes to be fairly simple and easy to predict. Such is not the case. In fact, just the opposite is true. The alkanes have greater variation in their reactivity than any other family. It is very difficult to restrict a reaction so that a particular product is obtained. Consider the reaction of propane with nitric acid at 500°C. If the products of the reaction were isolated, there would be four products. The following equations illustrate this reaction.

$-\overset{|}{\underset{|}{C}}-\overset{|}{\underset{|}{C}}-\overset{|}{\underset{|}{C}}-$ $\xrightarrow[500°C]{HNO_3}$

propane

(a) $-\overset{|}{\underset{|}{C}}-\overset{|}{\underset{|}{C}}-\overset{|}{\underset{|}{C}}-NO_2$

1-nitropropane

(b) $-\overset{|}{\underset{|}{C}}-\overset{|}{\underset{|}{C}}-\overset{|}{\underset{|}{C}}-$

         NO_2

2-nitropropane

(c) $-\overset{|}{\underset{|}{C}}-NO_2$

nitromethane

(d) $-\overset{|}{\underset{|}{C}}-\overset{|}{\underset{|}{C}}-NO_2$

nitroethane

four products

Notice that the four products formed include two reactions which involve the replacement of a hydrogen with a nitro group on the propane chain. However, two of the products (nitromethane and nitroethane) are formed by breaking a carbon bond in the original propane molecule and then the replacement of the hydrogen with a nitro group. Strong acids at high reaction temperatures often will have this effect on the alkane family.

Another reaction of interest is the reaction of a halogen with a member of the alkane family. To obtain a desired product, very limited restrictions must be placed on the reaction. As a simple example, consider the reaction of methane and chlorine gas.

First, let the reaction proceed uncontrolled. The products would be free carbon and hydrogen chloride in the gas phase. The reaction:

$$CH_4 + 2Cl_2 \xrightarrow{\text{(uncontrolled)}} C + 4HCl(g)$$

Now let us hold the temperature of the above reaction to 300°C and the pressure at 4.0 atm. If products are carefully removed and identified, a step-by-step reaction may be observed. Each step of the reaction will involve the substitution of one more chlorine atom to the alkane. It must be emphasized that this type of reaction is very difficult to control.

The following reactions illustrate this type of stepwise reaction for chlorine gas and methane.

Step 1: $CH_4 + Cl_2 \xrightarrow{300°C} CH_3Cl$ (chlormethane)

Step 2: $CH_3Cl + Cl_2 \xrightarrow{300°C} CH_2Cl_2$ (dichloromethane)

Step 3: $CH_2Cl_2 + Cl_2 \xrightarrow{300°C} CHCl_3$ (trichloromethane)

Step 4: $CHCl_3 + Cl_2 \xrightarrow{300°C} CCl_4$ (tetrachloromethane)

Uncontrolled: $CH_4 + 2Cl_2 \rightarrow C + 4HCl$ (carbon)

Many additional reactions for various families could also be considered. The reactions listed are but a small cross-section which will allow us to obtain a brief background on the chemical properties of the various organic families. In the next section we shall see how these family reactions can be incorporated in a full series of progressive reactions.

16.8 Progressive Reactions

A progressive reaction is a chemical reaction which begins with a reactant from one family and through a series of reactions, terminates with a product of another family. A typical reaction of this type would be to start with an alcohol and progress to an aldehyde and from an aldehyde to a carboxylic acid. As a specific example, consider the reaction of 1-propanol.

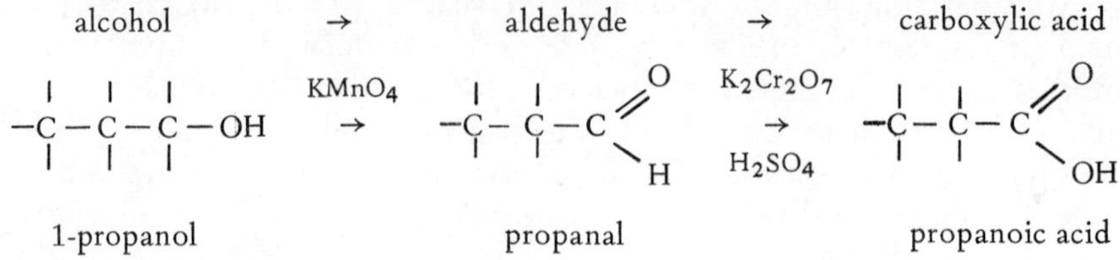

alcohol	→	aldehyde	→	carboxylic acid

$$-\overset{\displaystyle |}{\underset{\displaystyle |}{C}}-\overset{\displaystyle |}{\underset{\displaystyle |}{C}}-\overset{\displaystyle |}{\underset{\displaystyle |}{C}}-OH \quad \overset{KMnO_4}{\longrightarrow} \quad -\overset{\displaystyle |}{\underset{\displaystyle |}{C}}-\overset{\displaystyle |}{\underset{\displaystyle |}{C}}-C\overset{\displaystyle O}{\underset{\displaystyle H}{\diagup\!\!\!\diagdown}} \quad \overset{K_2Cr_2O_7}{\underset{H_2SO_4}{\longrightarrow}} \quad -\overset{\displaystyle |}{\underset{\displaystyle |}{C}}-\overset{\displaystyle |}{\underset{\displaystyle |}{C}}-C\overset{\displaystyle O}{\underset{\displaystyle OH}{\diagup\!\!\!\diagdown}}$$

1-propanol		propanal		propanoic acid

This progressive reaction involves the oxidation of the alcohol. Remember in the section dealing with the alcohol family reactions, how KMnO$_4$ or K$_2$Cr$_2$O$_7$ was used to oxidize the alcohol to the aldehyde. Further oxidation of the aldehyde yielded the appropriate carboxylic acid, in this case propanoic acid, which is also known as propionic acid.

Suppose a secondary alcohol were selected instead of a primary alcohol. A secondary alcohol is oxidized in one step to form a ketone. A ketone does not undergo further oxidation under normal conditions. Consider the reaction of 2-butanol to illustrate this reaction.

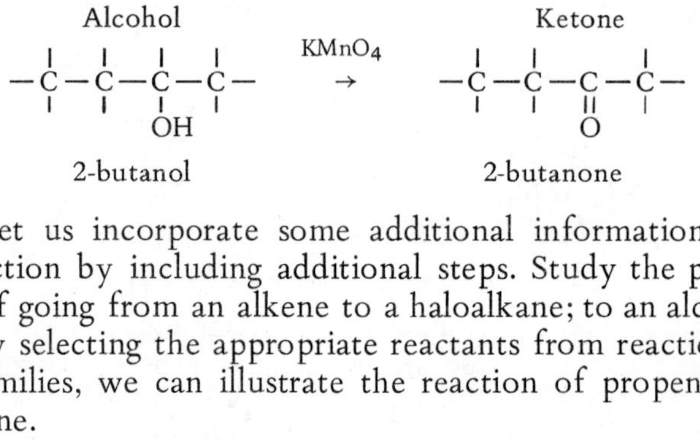

Alcohol		Ketone

$$-\overset{|}{\underset{|}{C}}-\overset{|}{\underset{|}{C}}-\overset{|}{\underset{OH}{C}}-\overset{|}{\underset{|}{C}}- \quad \overset{KMnO_4}{\longrightarrow} \quad -\overset{|}{\underset{|}{C}}-\overset{|}{\underset{|}{C}}-\overset{}{\underset{O}{C}}-\overset{|}{\underset{|}{C}}-$$

2-butanol		2-butanone

Now let us incorporate some additional information into the above reaction by including additional steps. Study the progressive reaction of going from an alkene to a haloalkane; to an alcohol, to a ketone. By selecting the appropriate reactants from reactions of the various families, we can illustrate the reaction of propene to form 2-propanone.

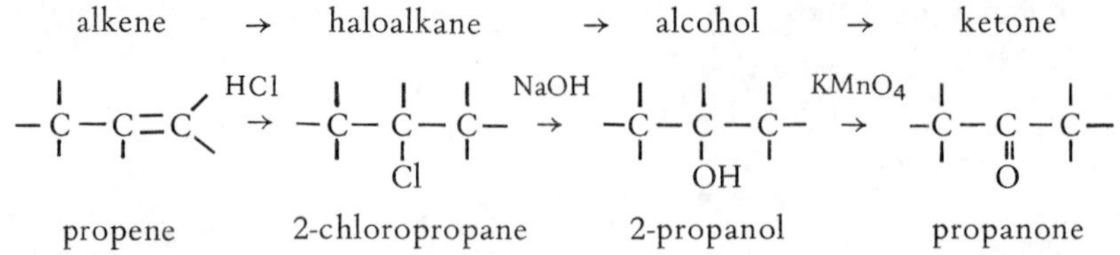

alkene	→	haloalkane	→	alcohol	→	ketone

$$-\overset{|}{\underset{|}{C}}-C\!\!\equiv\!\!C\diagdown \quad \overset{HCl}{\longrightarrow} \quad -\overset{|}{\underset{|}{C}}-\overset{|}{\underset{Cl}{C}}-\overset{|}{\underset{|}{C}}- \quad \overset{NaOH}{\longrightarrow} \quad -\overset{|}{\underset{|}{C}}-\overset{|}{\underset{OH}{C}}-\overset{|}{\underset{|}{C}}- \quad \overset{KMnO_4}{\longrightarrow} \quad -\overset{|}{\underset{|}{C}}-\overset{}{\underset{O}{C}}-\overset{|}{\underset{|}{C}}-$$

propene	2-chloropropane	2-propanol	propanone

Let us consider one additional progression which is often illustrated in organic chemistry laboratories. This is the progression from an alkane to a haloalkane, to an alcohol, to an aldehyde, to a carboxylic acid. Taking a specific example, consider the reaction of propane to eventually form propanoic acid. The portion of this reaction from the alcohol to the carboxylic acid has been discussed earlier in this section.

Alkane	Haloalkane	Alcohol	Aldehyde	Carboxylic Acid

$$-\overset{|}{\underset{|}{C}}-\overset{|}{\underset{|}{C}}-\overset{|}{\underset{|}{C}}-\ \underset{300^\circ C}{\overset{Cl_2}{\longrightarrow}}\ -\overset{|}{\underset{|}{C}}-\overset{|}{\underset{|}{C}}-\overset{|}{\underset{|}{C}}-Cl\ \overset{NaOH}{\longrightarrow}\ -\overset{|}{\underset{|}{C}}-\overset{|}{\underset{|}{C}}-\overset{|}{\underset{|}{C}}-OH\ \overset{KMnO_4}{\longrightarrow}\ -\overset{|}{\underset{|}{C}}-\overset{|}{\underset{|}{C}}-\overset{\nearrow O}{\underset{\searrow H}{C}}\ \underset{H_2SO_4}{\overset{K_2Cr_2O_7}{\longrightarrow}}\ -\overset{|}{\underset{|}{C}}-\overset{|}{\underset{|}{C}}-\overset{\nearrow O}{\underset{\searrow OH}{C}}$$

propane	1-chloropropane	1-propanol	propanal	propanoic acid

From these brief examples you can see how a chemist can form a series of organic chemical families from a different family through a series of chemical reactions. One distinct advantage of organic chemistry is that each member of a given family generally reacts in the same manner as other members of that family. Thus, if you learn what reaction will convert one species to another, it will generally hold true for all cases.

Organic chemistry is a vast area and we have but scratched the surface. It is hoped, however, that this brief discussion will suggest further study to the student who continues in chemistry. To the terminal student, it is hoped that these chapters have given you sufficient background to recognize, name, and realize general properties of organic substances as they are found in daily life.

Glossary

Bond Angle—A bonding parameter. The bond angle is the number of degrees in an angle formed by the central atom and two adjacent atoms.

Bond Length—A bonding parameter. The distance from the center of one bonded atom to the center of the other atom participating in the bond. This distance is usually expressed in Ångstroms.

Isomerism—The existence of two or more chemical species with the same molecular formula and molecular mass but not the same physical properties. Types of isomerism include positional, skeletal, functional, and optical.

Markownikoff's Rule—In the addition of an acid, HX, to an unsymmetric alkene, the hydrogen from the acid attaches to that carbon atom of the double bond which already has the greater number of hydrogen atoms attached to it. The X group attaches to the other carbon atom of the double bond.

Pi Bond (π)—A molecular bond formed when bonding electrons are shared in such a manner that their electron distributions lie above and below the axis of the sigma bond.

Sigma Bond (σ)—A molecular bond formed when bonding electrons are shared in such a manner that their electron distributions lie along the axis of the molecule.

Exercises

1. Draw structures for all isomers with the chemical formula $C_5 H_{12}$. Give the correct I.U.P.A.C. name in each case.

2. Given the formula $C_4 H_{10} O$, illustrate positional, skeletal, functional, and optical isomerism by selecting pairs of structures.

3. Butane is reacted with nitric acid at 500°C. Draw structures and write the correct I.U.P.A.C. name for all possible products. (Hint–there are at least six.)

4. Starting with ethene, write progressive reactions to illustrate the production of (a) a haloalkane, (b) an alcohol, (c) an aldehyde, (d) a carboxylic acid. List the reactants and name all products with the correct I.U.P.A.C. name.

5. Indicate the organic product formed in the following reactions.
   a.   Ethanol + sodium hydroxide =
   b.   Phenol + sodium metal =
   c.   Methanol + sodium metal =
   d.   Phenol + sodium hydroxide =

6. Starting with 2-pentanol, write an equation for the formation of 2-pentanone.

7. Complete the following chemical equations.
   a. $CH_3 OH + K_2 Cr_2 O_7 \rightarrow$
   b. $CH_3 CHOHCH_3 + KMnO_4 \rightarrow$
   c. $(CH_3)_3 COH + KMnO_4 \rightarrow$

8. Complete the following chemical equations.
   a. $CH_3 CHO + K_2 Cr_2 O_7 \xrightarrow{H_2 SO_4}$
   b. $CH_2 CH_2 + HCl \rightarrow$
   c. $CH_3 CH_2 CH_2 Cl + NaOH \rightarrow$

9. Starting with butane, write a series of reactions to illustrate the formation of:
   a.   A haloalkane   c.   An aldehyde
   b.   An alcohol   d.   A carboxylic acid

10. Starting with 2-butene, write a series of reactions to illustrate the formation of:
   a.   A haloalkane   c.   A ketone
   b.   An alcohol

CHAPTER 17

Biochemistry—Carbohydrates

17.1 Introduction

An area of chemistry that has become very popular in the past few years is biochemistry. Possibly because of its relationship with life and more specifically humans, biochemistry has become increasingly important in today's world. The key to many of mankind's ecological improvements lies in the area of biochemistry. Biochemistry is a sophisticated branch of chemistry. The chemical species are generally quite complex, and this would seem logical when we realize that something as complex as the human body is involved.

As in the study of organic chemistry, an overview will be given of important families, their nomenclature, and their reactivity. It is hoped that this preliminary information will enable the interested student to pursue this area in future study.

17.2 Carbohydrates

Of all the different classes of compounds which make up biochemistry, the carbohydrates are the largest in number and thus of prime consideration. Carbohydrates include three subgroups: the sugars, the starches, and cellulose. The sugars are ready energy for a living organism, whereas the starches are energy forms which must undergo chemical change in order to be assimilated by a living organism. Let us first consider the sub-group of sugars with respect to structure and nomenclature.

17.3 Sugars

The sugars can be divided into three classes depending on their structure. The first class is called the *monosaccharides,* or simple sugars; the second class is called the *disaccharides,* or double sugars; and the third class which involves very complex polymers is called the *polysaccharides.* Starches and cellulose would be examples of polysaccharides.

Monosaccharides can be further classified if one were to look at such factors as the number of carbon atoms and the detailed structure of the molecule. If a monosaccharide contains three carbon atoms, it is called a *triose,* four carbon atoms a *tetrose,* five carbon atoms a *pentose,* and six carbon atoms a *hexose.* Also, if the sugar contains an aldehyde group, it is called an *aldose,* and if it contains a keto group, it is called a *ketose.* Notice that all sugars end in the suffix -ose.

Of the various monosaccharides, there are two of major importance that we shall consider. They are called glucose and fructose.

Glucose is the most abundant monosaccharide. It is also the basic unit for the more complex carbohydrates such as starch and cellulose. A common name for glucose is dextrose. Glucose can be classified as an aldohexose. This means that it is a six carbon sugar with an aldehyde group somewhere in the structure. The molecular formula for glucose is $C_6H_{12}O_6$. The structure is illustrated in Figure 17.1.

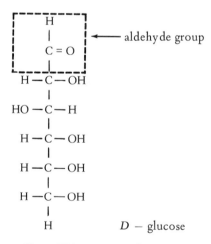

Figure 17.1. Structure of Glucose, $C_6H_{12}O_6$.

Notice the double bonded oxygen and singly bonded hydrogen on the top carbon atom. This is the aldehyde group which makes this an aldose sugar. The *D*- before the name of glucose relates to isomerism and will be discussed in Section 17.4.

The other monosaccharide of interest is *fructose*. Fructose is found in many fruits as well as honey. In terms of structure, fructose is considered a ketose and because it contains six carbon atoms, it is called a ketohexose. The chemical formula for fructose is $C_6H_{12}O_6$, as is glucose. The positioning of the double bonded oxygen is the structural difference between glucose and fructose. The structure of fructose is given in Figure 17.2.

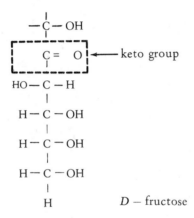

Figure 17.2. Structure of Fructose, $C_6H_{12}O_6$.

In terms of chemical reactivity, glucose and fructose combine together to form the disaccharide sucrose. *Sucrose*, $C_{12}H_{22}O_{11}$, is commonly known as table sugar. Sucrose is derived commercially from either sugar beets or sugarcane. However, it is also found in the sap of many other plants. In terms of structure, sucrose becomes a fairly complex molecule. It is difficult to illustrate sucrose in two dimensions as you must envision the molecule in three dimensions in order to attain the proper perspective. Figure 17.3 illustrates the structure of sucrose.

In Section 17.6, various reactions of sucrose will be considered; one reaction is the hydrolysis of sucrose to give glucose and fructose. It is this reaction which led to the identification of the sucrose structure.

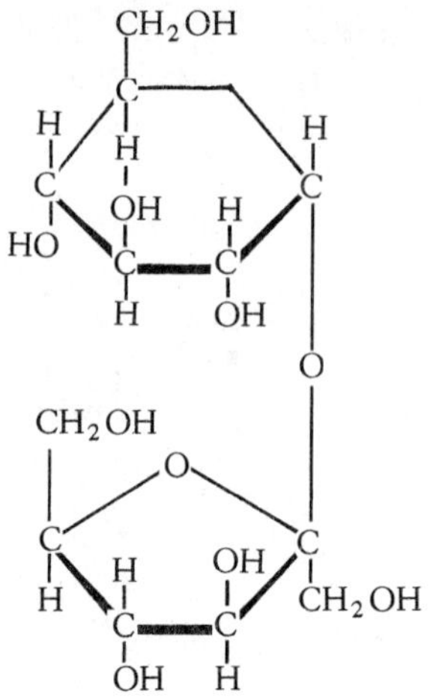

Figure 17.3. Structure of Sucrose, $C_{12}H_{22}O_{11}$.

After the disaccharides, the next group of carbohydrates according to structure, is the polysaccharides. The polysaccharides are composed of many monosaccharide units in each molecule. In fact, there may be hundreds or thousands of monosaccharide units in one polysaccharide molecule.

17.4 Isomerism

The general definition of isomerism and its function in organic compounds was discussed in Chapter Sixteen. Of the various types of isomerism it is optical isomerism that is of importance when discussing sugars. Before a discussion of reactivity of the various compounds can be made, an understanding of optical isomerism is necessary.

When discussing a sugar, the term "active" is often used. That is, we say we have an "active sugar." This means that the molecule is active optically. To understand what optically active means, you first must understand light. Light normally vibrates in all directions equally; however, we can create *polarized light,* or light that vibrates in but one direction. It is the polarized light that we are concerned with in optical isomerism. See Figure 17.4.

An optically active compound is one that rotates the plane of polarized light. In other words, if a beam of polarized light were to pass through an optically active substance, the beam would be at a different angle when it emerges from the substance. By instrumenta-

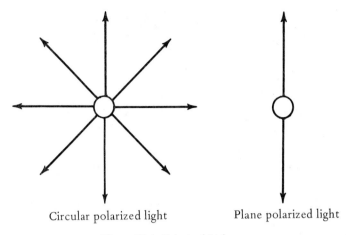

Circular polarized light   Plane polarized light

Figure 17.4. Polarized Light.

tion, this angle may be measured and a definite number of degrees of rotation obtained. If the beam of polarized light is rotated in a clockwise direction, the substance is said to be *dextrorotatory*. On the other hand, if the beam of polarized light is rotated in a counterclockwise direction, the substance is said to be *levorotatory*. The symbols (+) and (−) are used to indicate *dextro*rotatory and *levoro*tatory respectively. A method of indicating the structure is the use of the symbols -*D*- and -*L*-.

To return to isomerism, the definition of optical isomerism was when two structures were nonsuperimposable mirror images of each other. Now, however, greater clarification must be made. A compound is optically active if its molecules are not superimposable on their mirror images. A compound is optically inactive if its molecules are superimposable on their mirror images. Thus, there are two types of optical isomerism which we must be concerned with.

Of the optically active isomers, the form (+)-glucose is of interest with respect to the monosaccharides. (+)-glucose undergoes a completely different set of reactions than a (−)-hexose, and for this reason, it is critical to indicate which isomer is being discussed.

17.5 Reactions of (+)-Glucose

The production of (+)-glucose from carbon dioxide and water in plants is called *photosynthesis*. The catalyst for this reaction in plants is chlorophyll and energy is also required for the reaction to occur. The energy is usually provided by sunlight. In equation form:

$$6CO_2 + 6H_2O \xrightarrow[\text{cholorophyll}]{\text{energy}} C_6H_{12}O_6 + 6O_2$$

This is the basic equation by which life is sustained on earth. Without the production of oxygen and glucose, life as we know it could not exist.

Since (+)-glucose is the most common sugar we have, it is of interest to note some of the typical reactions which it undergoes. In this way, an idea of typical monosaccharide reactions can be obtained.

(+)-glucose reacts with nitric acid, HNO_3, to form glucaric acid (saccharic acid). This reaction is written:

CHO		COOH
CHOH		CHOH
CHOH	HNO_3 $\rightarrow$	CHOH
CHOH		CHOH
CHOH		CHOH
CH_2OH		COOH
(+)-glucose		glucaric acid

Notice that the four middle carbons are unchanged in the reaction. It is only the end carbons which achieve the carboxylic acid structure.

Now consider the reaction of (+)-glucose with hydrogen gas using a nickel catalyst. The reaction is:

CHO		CH_2OH
CHOH		CHOH
CHOH	H_2 $\rightarrow$ Ni	CHOH
CHOH		CHOH
CHOH		CHOH
CH_2OH		CH_2OH
(+)-glucose		glucitol (sorbitol)

Again, all changes in structure occur on the end carbons, and not at the middle four carbons. However, if a strong acid such as HI is added to glucitol and the system heated, the latice chain becomes vulnerable. This is illustrated in the following reaction:

$$
\begin{array}{ccc}
\text{CH}_2\text{OH} & & \text{CH}_3 \\
| & & | \\
\text{CHOH} & & \text{CHI} \\
| & & | \\
\text{CHOH} & \text{HI} & \text{CH}_2 \\
| & \rightarrow & | \\
\text{CHOH} & \text{heat} & \text{CH}_2 \\
| & & | \\
\text{CHOH} & & \text{CH}_2 \\
| & & | \\
\text{CH}_2\text{OH} & & \text{CH}_3 \\
\text{glucitol} & & \text{2-iodohexane}
\end{array}
$$

Thus, in a two-step reaction, you have gone from the realm of a carbohydrate to the familiar family of haloalkanes. This should help emphasize the close relationship among the various families in organic and biochemistry.

In organic and biochemistry there are certain reagents which are used to identify the various chemical families. For example, a solution called *Fehling's* solution is used to identify an aldehyde group. Fehling's solution is composed of a copper(II) ion complexed with a tartrate ion. When Fehling's solution oxidizes an aldehyde group, a red precipitate is obtained. This is one method of identifying a reducing sugar, as the sugar contains the aldehyde group. Another reagent which will also identify a sugar is called *Benedict's* solution. Benedict's solution is a complex between the copper(II) ion and the citrate ion which also oxidizes the aldehyde group to give a red precipitate. This reaction can be illustrated by the equation:

$$
\begin{array}{ccc}
\text{RCHO + Benedict's} & \rightarrow & \text{RCOO}^- + \text{Cu}_2\text{O}(s) \\
\text{solution} & & \text{(Red} \\
\text{(blue} & & \text{ppt.)} \\
\text{solution)} & &
\end{array}
$$

The drawback is that these reagents would show the same reaction with many reducing groups and thus have limited application to carbohydrate, starch, and protein systems.

17.6 Reactions of Disaccharides

Earlier in this chapter, it was noted that sucrose was formed by combining (+)-glucose and (−)-fructose. When sucrose is hydrolyzed, the two monosaccharides are obtained. A unique property of sucrose is that it is a non-reducing sugar. Thus, sucrose will not reduce Tollen's or Benedict's solution. This leads to the conclusion that sucrose does not contain a "free" aldehyde group. If you were to study the structure of sucrose, you would notice that there are no free aldehyde groups in the structure.

In equation form, the hydrolysis of sucrose to form (+)-glucose and (−)-fructose is shown in Figure 17.5.

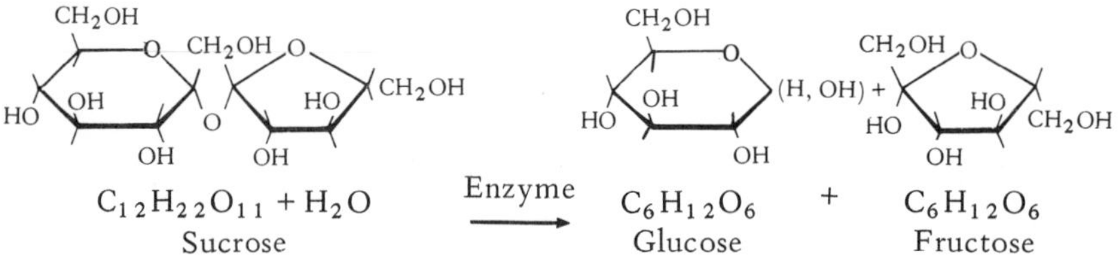

$$C_{12}H_{22}O_{11} + H_2O \quad \xrightarrow{\text{Enzyme}} \quad C_6H_{12}O_6 \quad + \quad C_6H_{12}O_6$$

Sucrose   Glucose   Fructose

Figure 17.5. The Decomposition of Sucrose.

17.7 Cellulose and Starch

The two most important polysaccharides are cellulose and starch. Both cellulose and starch are produced in plants by the process of photosynthesis. In terms of structure, both are composed of glucose units.

The shape and size of the *starch* granule is dependent on the plant from which it is obtained. Some common plants from which starch is obtained are corn, wheat, barley, rice, potatoes, and other grain-type crops. Starch is essentially insoluble in cold water; but if put in hot water, the starch granules will swell to such an extent that the granule bursts.

In terms of composition, starch is essentially composed of two carbohydrates. Approximately 20% of starch is *amylose* and 80% is *amylopectin*. Both of these carbohydrates have the same general formula $(C_6H_{10}O_5)_n$. The value of n will vary among the carbohydrates, but is a very larger integer. To gain a little insight into the structure of amylose and amylopectin, it should be remembered that both of these carbohydrates are composed of a series of glucose units. Figure 17.6 illustrates a partial picture of the amylose structure.

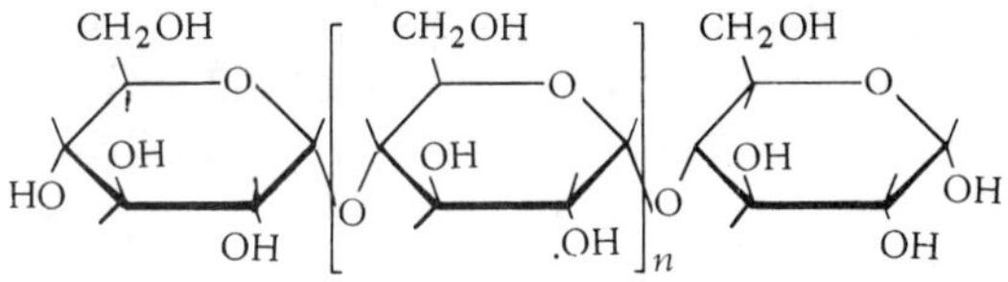

Starch (amylose, n = 100—1000)

Figure 17.6. A Portion of an Amylose Molecule.

Experimentation has shown that there are about 200 units of glucose per molecule of amylose.

Amylopectin, on the other hand, has more than 1,000 glucose units per molecule. The structure for amylopectin is more complex than amylose. It is difficult to illustrate the structure of amylopectin but Figure 17.7 shows a portion of an amylopectin molecule.

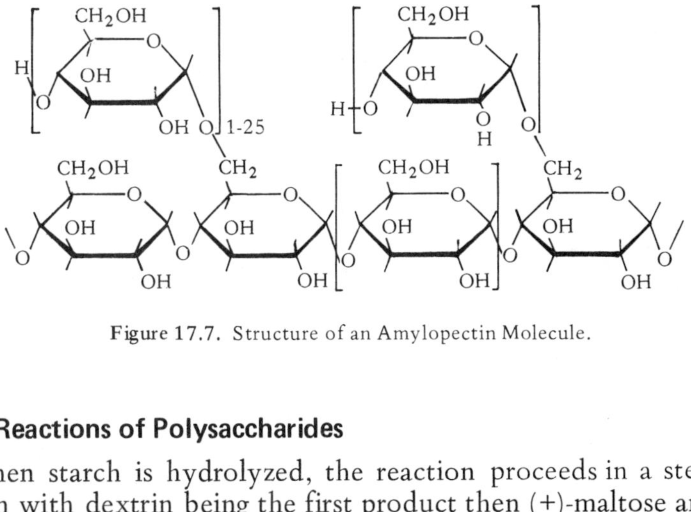

Figure 17.7. Structure of an Amylopectin Molecule.

17.8 Reactions of Polysaccharides

When starch is hydrolyzed, the reaction proceeds in a stepwise fashion with dextrin being the first product then (+)-maltose and the final product is (+)-glucose. This reaction is illustrated in Figure 17.8.

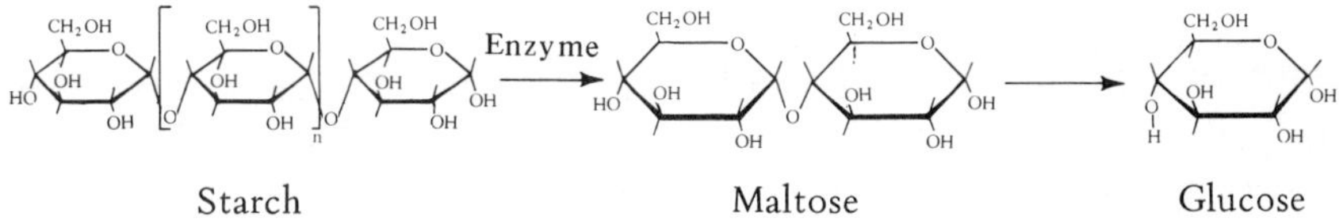

Starch   Maltose   Glucose

Figure 17.8. The Hydrolysis of Starch.

The hydrolysis of cellulose also produces (+)-glucose. However, the linkage between the glucose units in cellulose is different than in starch. Thus, the total structure of cellulose varies from starch by the orientation or linking together of the numerous glucose units. Therefore, cellulose will have different reactions from starch. Cellulose is used as the basis for many artificial fabrics such as rayon.

Of the various reactions of carbohydrates which are studied, the most important would involve the metabolism of carbohydrates in the human body. Simple sugars such as (+)-glucose can be readily absorbed by the human body. Other carbohydrates such as starch must be hydrolyzed to the disaccharide and further to the monosaccharide for absorption by the body. The process of digestion begins in the mouth with the hydrolysis of starches to disaccharides such as maltose. The disaccharides are converted to monosaccharides such as glucose in the stomach and intestinal tract. From the intestinal tract, the monosaccharides pass into the blood stream and are absorbed by the cells. Figure 17.9 illustrates the digestive process for various carbohydrates. Notice the key role that the liver plays in digestion. It is in the liver that all carbohydrates other than glucose are finally converted to glucose for absorption by the body.

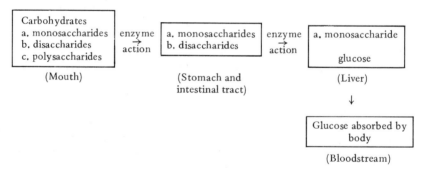

Figure 17.9. Digestion and Absorption of Carbohydrates.

The other polysaccharide of interest is *cellulose*. Cellulose is found in plant fibers and wood. Such substances as cotton and wood are nearly all cellulose. Cellulose is composed of at least 10,000 glucose units and is insoluble in water. As was the case with starch, cellulose has the general formula $(C_6H_{10}O_5)_n$. The basic difference between cellulose and starch is the structure of the molecule. Again, due to the size of the molecule, only a portion of a cellulose molecule may be illustrated. Figure 17.10 illustrates a portion of the cellulose molecule. Compare its structure with that of amylose and amylopectin and notice the difference in the linkage.

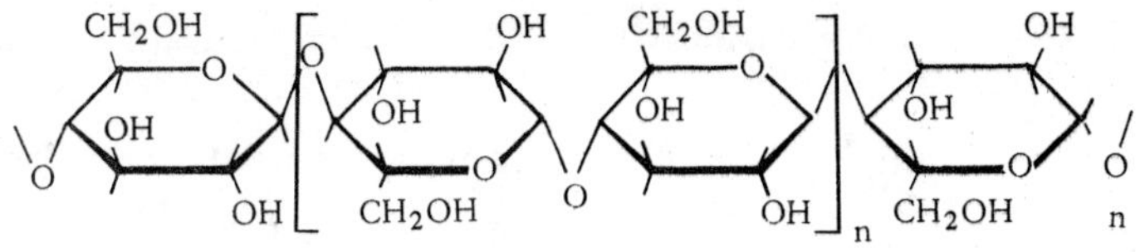

Cellulose

n ≈ 10,000

Figure 17.10. Structure of a Cellulose Molecule.

Glossary

Benedict's Solution—A complex solution of the copper (II) ion and the citrate ion used to test for the aldehyde group. Used to confirm the presence of a reducing sugar in a solution.

Carbohydrate—A group of organic compounds composed of carbon, hydrogen, and oxygen. Sugars, starches, and cellulose are all members of this group.

Cellulose—A carbohydrate found in cell walls and woody parts of plants. Cellulose has the general formula $(C_6H_{10}O_5)n$.

Dextrorotatory—The ability of a substance to rotate the plane of polarization in a clockwise direction.

Disaccharide—A double sugar with the formula $C_{12}H_{22}O_{11}$. Examples would be sucrose, maltose, and lactose.

Fehling's Solution—A complex copper(II) ion and tartrate ion solution used to confirm the presence of an aldehyde group in a compound. Gives a red precipitate upon reaction with an aldehyde group.

Hexose—Any sugar which contains six carbon atoms.

Levorotatory—The ability of a substance to rotate the plane of polarization in a counter-clockwise direction.

Monosaccharide—A simple sugar. A sugar which cannot be broken down by hydrolysis. Glucose and fructose are examples of monosaccharides.

Photosynthesis—The chemical process of producing carbohydrates and oxygen from carbon dioxide and water in plants.

Polysaccharide—A complex carbohydrate which decomposes by hydrolysis to monosaccharides. Examples of polysaccharides would be starch and cellulose.

Starch—A polysaccharide which contains many glucose units. Starches have the general formula $(C_6H_{10}O_5)n$.

Exercises

1. Classify each of the following as a monosaccharide, disaccharide, or polysaccharide and draw the molecular structure.

   a. Glucose   c. Fructose

   b. Sucrose   d. Corn starch

2. Explain the difference between the two sugars, glucose and fructose.

3. Discuss why starch is not a food form for immediate energy.

4. What is the difference between a keto-sugar and an aldo-sugar?

5. Write an equation for the hydrolysis of sucrose.

6. Describe the basic difference between cellulose and starch.

7. Draw structures for the following substances:

a.	(+)-glucose	d.	Cellulose
b.	(−)-fructose	e.	Starch
c.	Sucrose		

8. Explain how you would distinguish between a solution of (+)-glucose and a solution of corn starch.

9. What is the basic difference between cellulose and starch?

10. Assuming that the process of photosynthesis procedes with 100% efficiency, determine the number of grams of glucose, $C_6H_{12}O_6$, that can be produced when 440 grams of carbon dioxide reacts with sufficient water.

11. Draw structures and predict the product when (+)-fructose reacts with H_2 and a nickel catalyst at high temperatures and pressure.

12. Describe the total process of metabolism from the time food is taken into the mouth until it is absorbed by the cells.

13. Draw and label all structures in the hydrolysis of starch to glucose.

CHAPTER 18

Biochemistry—Fats and Proteins

18.1 Common Fats and Their Chemical Names

When a person refers to a substance as a fat, the chemist realizes he is referring to a chemical family called *esters*. An ester is a functional derivative of the carboxylic acids and in general, has the structure

$$R - C \overset{\displaystyle O}{\underset{\displaystyle OR'}{\diagdown}}$$

R and R' are alkyl or aryl groups.

To clarify, an ester is formed when a carboxylic acid reacts with an alcohol. For example, consider the reaction of acetic acid and 1-butanol.

$$CH_3COOH + CH_3CH_2CH_2CHOH \rightleftharpoons CH_3COOCHCH_2CH_2CH_3$$

(acetic acid)   (1-butanol)   (*n*-butyl ethanoate)

$$-\overset{|}{\underset{|}{C}}-C\overset{\displaystyle O}{\underset{\displaystyle O\ \fbox{H}}{\diagup}} + -\overset{|}{\underset{|}{C}}-\overset{|}{\underset{|}{C}}-\overset{|}{\underset{|}{C}}-\overset{|}{\underset{|}{C}}-\fbox{OH} = -\overset{|}{\underset{|}{C}}-\overset{\displaystyle O}{\overset{||}{C}}-\overset{}{O}-\overset{|}{\underset{|}{C}}-\overset{|}{\underset{|}{C}}-\overset{|}{\underset{|}{C}}-\overset{|}{\underset{|}{C}}-$$

This is a classic reaction and one which illustrates the activity of the bond positions of both the carboxylic acid and the alcohol. Most esters are formed in this manner.

In general usage, the common esters are either animal or vegetable *fats*. If the fat is in a liquid form, it is referred to as an *oil*. Some common sources of fats or oils would be corn oil, coconut oil, peanut oil, cottonseed oil, butter, lard and tallow.

Biologically, fats cannot be absorbed in their entirety as a form of food and energy. Rather, they must be broken down into simpler forms for use by the body.

Table 18.1 lists some of the common fats; their chemical structure, and name.

TABLE 18.1
COMMON FATS

Fatty Acid	Formula	Source
butyric acid	C_3H_7COOH	butter fat
caproic acid	$C_5H_{11}COOH$	butter fat
caprylic acid	$C_7H_{15}COOH$	coconut oil
capric acid	C_9H_9COOH	coconut oil
lauric acid	$C_{11}H_{23}COOH$	coconut oil
myristic acid	$C_{13}H_{27}COOH$	nutmeg oil
arachidic acid	$C_{19}H_{39}COOH$	peanut oil
palmitic acid	$C_{15}H_{31}COOH$	palm oil
stearic acid	$C_{17}H_{35}COOH$	tallow
uleic acid	$C_{17}H_{33}COOH$	olive oil
linoleic acid	$C_{17}H_{31}COOH$	cottonseed oil
linolenic acid	$C_{17}H_{29}COOH$	linseed oil
ricinoleic acid	$C_{17}H_{32}OHCOOH$	castor oil

Of the various food groups, fats are of major importance because of their ability to function as a reserve energy system as well as a source of insulation and lubrication in the body. One drawback is that an excess of fats can cause obesity and heart disease which is why many older people are concerned with low-fat diets.

18.2 Reactions of Fats (Reduction)

As was mentioned earlier, fats are esters. Therefore, when we discuss the reactivity of fats, we can discuss the reactivity of esters.

Esters can be reduced by two methods. The first is by catalystic hydrogenation using molecular hydrogen. The second method is by chemical reduction using alcohol and a metal such as sodium. Figure 18.1 illustrates these two processes of reduction.

General:	$R-\overset{O}{\overset{\|\|}{C}}-O-R'$ ester	$\xrightarrow{\text{reduction}}$	RCH_2OH alcohol	+	$R'OH$ alcohol
Case 1:	$C_3H_2COOC_2H_5$ ethyl butanoate	$\underset{\text{cat}}{\overset{H_2}{\rightarrow}}$	$C_3H_7CH_2OH$ butanol	+	C_2H_5OH ethanol
Case 2:	$C_5H_{11}COOC_2H_5$ ethyl hexanoate	$\underset{C_2H_5OH}{\overset{Na}{\rightarrow}}$	$C_5H_{11}CH_2OH$ hexanol	+	C_2H_5OH ethanol

Figure 18.1. Reduction of Fats.

When discussing fats, the term *"glycerides"* is commonly used. A glyceride is also a fat or an ester. The name glyceride is obtained by noting the alcohol from which the ester was produced. The alcohol, $HOCH_2CHOHCH_2OH$, is called *glycerol* and thus any ester obtained from glycerol has the name glyceride. The hydrolysis of glycerides is used for the production of soaps. In general, when a glyceride is hydrolyzed with a base, the two products formed are glycerol and soap. Figure 18.2 shows the general reaction.

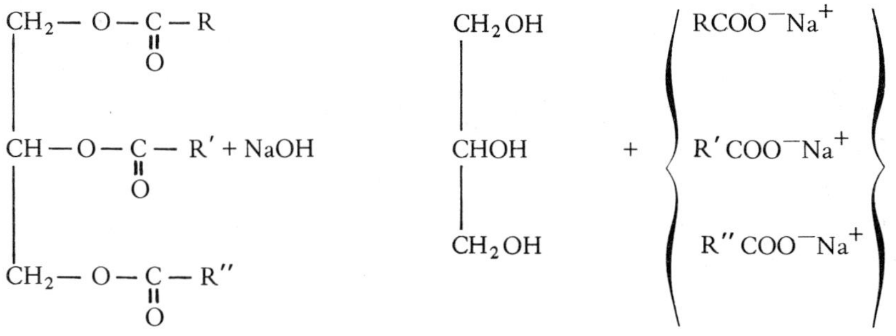

a glyceride glycerol soap

Figure 18.2. Saponification of an Ester.

The most common soaps which are marketed today are a mixture of sodium salts and fatty acids. In addition to sodium, potassium is also used in the production of soap. A potassium soap is a soft soap and is more commonly used in a liquid soap.

It should be emphasized that soaps are detergents. Even though detergents act essentially the same way as soaps, they are markedly different in their synthesis. Many detergents are made from alcohols obtained from some fat such as coconut oil. Most detergent molecules have a large non-polar hydrocarbon end that is oil-soluble and a polar end that is water-soluble. Detergents are salts of strong acids or strong bases whereas soaps are salts of weak acids. In today's world, much concern about detergents and our environment is being expressed. The reason for this is that the entire detergent molecule is not water soluble and as a result, there is a pollution problem of detergents in our rivers and streams. Soaps do not cause this problem because they are destroyed by bacteria. To cure this ecological problem, a different method of disposal of detergent wastes must be used. We no longer can merely dump them in our rivers and expect nature to eliminate them.

18.3 The Proteins

The family of proteins can be divided into two classes based on their molecular shape or structure. The first class is *fibrous proteins* which are long and threadlike. Fibrous proteins are insoluble in water. The second class is called *globular proteins,* which are folded in small units that are somewhat spherical in shape. Globular proteins are soluble in water.

The major use of fibrous proteins is structure building in the body. Thus the long, striated, rigid structure is a beneficial characteristic. Examples of some fibrous proteins would be *keratin* in hair or nails, *collagen* in tendons, *myosin* in muscle, and *fibroin* in silk.

Many globular proteins are involved in the normal body processes. Globular proteins are found in enzymes, hormones, and antibodies. Some common globular proteins would be *albumin* in eggs, *hemoglobin* in blood, *insulin* from the pancreas, *thyroglobulin* from the thyroid gland and ACTH from the adrenal cortex.

After proteins are divided according to solubility, the structure becomes of major interest as well. There are two levels of study for the structure. The first is the *primary* level, or in other words, the way the atoms of the protein molecules arrange themselves in various bonding chains. The second level of study is referred to as the *secondary* level and involves the spacial structure of the total molecule.

That is, whether the molecule is a coil, a strip, a sheet or a spherical shape. A more sophisticated study would include additional levels but these two will be sufficient for our use.

Before discussing the primary structure of proteins, it must be understood that proteins are polymers of amino acids. That is, proteins are hydrolyzed by strong acids to form amino acids. When we discuss proteins then, we are actually referring to amino acid residues connected together by amide groups.

In general, the most important method of combining amino acids is the *peptide* bond. The reason for this is that proteins are composed of a series of peptide bonds. At the primary level of structure, we have the peptide bonds. This is illustrated in Figure 18.3.

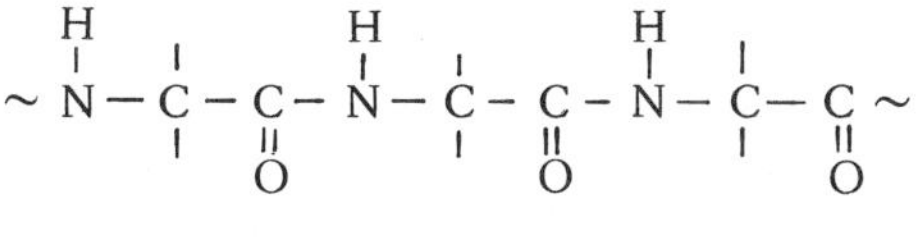

Figure 18.3. A Portion of a Peptide Chain.

All proteins contain this basic peptide chain. However, to every third atom of the chain will be a side chain. This side chain varies with the given amino acid involved. The basic peptide chain becomes more complex as illustrated in Figure 18.4. The various R groups indicate the attached side groups.

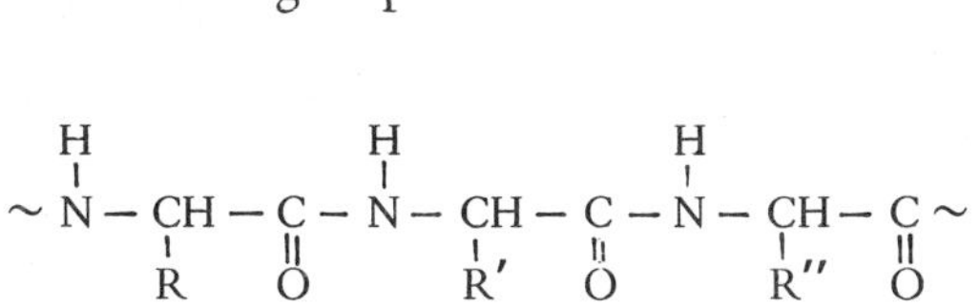

Figure 18.4. Sidechain Structure of Peptide Chain.

The side chains affect such properties as acidity, size, shape, and electronegativity of the molecule.

The primary level involves the straight chain as well as the attached side chains. Now let us consider the secondary levels.

From the primary level we know the structure of proteins involves a series of peptide units and functional side chains. The question now becomes how they orient themselves in space. Are they straight, threadlike, in a coil, or in a spherical shape. Through experimental methods such as x-ray patterns, the various shapes of the proteins have been determined. The secondary level is very complex and we shall be content to observe the result of a few common proteins to realize the complexity of the problem. Figures 18.5 and 18.6 illustrate the pleated, sheet structure of fibroin and the helix structure of keratin.

18.4 Reactions of Proteins and Amino Acids

The structure for amino acids varies greatly. There may be straight-chained structures or there can be cyclic structures. All amino acids do have some general properties in common. First, amino acids are non-volatile crystalline solids which melt at high temperatures. Second, amino acids are insoluble in non-polar solvents like benzene or ether. They are soluble in water. The K_A or K_B constants are low, indicating very little dissociation.

All amino acids other than glycine have at least one asymmetric carbon atom. Therefore, every amino acid other than glycine is optically active. In terms of reactivity, amino acids are like any compound containing an amino or a carboxyl group.

As was mentioned earlier, the key to protein structure is the peptide linkage. *Peptides* are amides formed by the combining of carboxyl and amino groups of amino acids. The amide group— $-\underset{\underset{H}{|}}{N}-\underset{\underset{O}{||}}{C}-$ is referred to as the *peptide linkage.* Just as some carbo-hydrates are compounds of simple sugars, the amino acids link together to form multiple peptide units within a protein molecule. Such terms as dipeptides, tripeptides, and polypeptides are used to describe the number of linkages per molecule.

As an overview, it is of interest to study the structure of some common amino acids found in protein. There are others but the amino acids in Figure 18.7 are commonly found in man's everyday diet.

18.5 The DNA and RNA Molecules

As a final topic in biochemistry, let us observe the *nucleoproteins,* or in other words, those substances produced when proteins combine with another polymer called a *nucleic* acid. These nucleic acids are now known to be the basis of heredity.

The structural basis of any nucleic acid is a *polynucleotide* chain. This corresponds to the peptide chain in protein. The polynucleotide chain is composed of phosphoric acid, sugar, and nitrogen base units called nucleotides. The structure of a polynucleotide chain is shown in Figure 18.8.

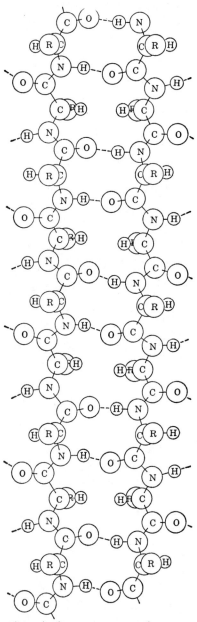

Pleated sheet structure (*beta arrangement*) proposed by Pauling for silk fibroin.

Figure 18.5. Structure of Silk Fibroin.

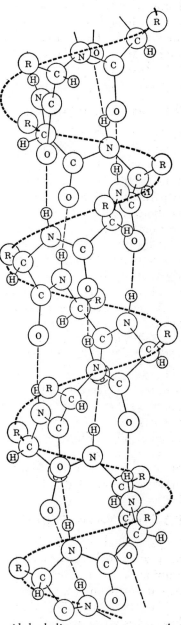

Alpha helix structure proposed by Pauling for α–keratin.

Figure 18.6. Structure of α–keratin.

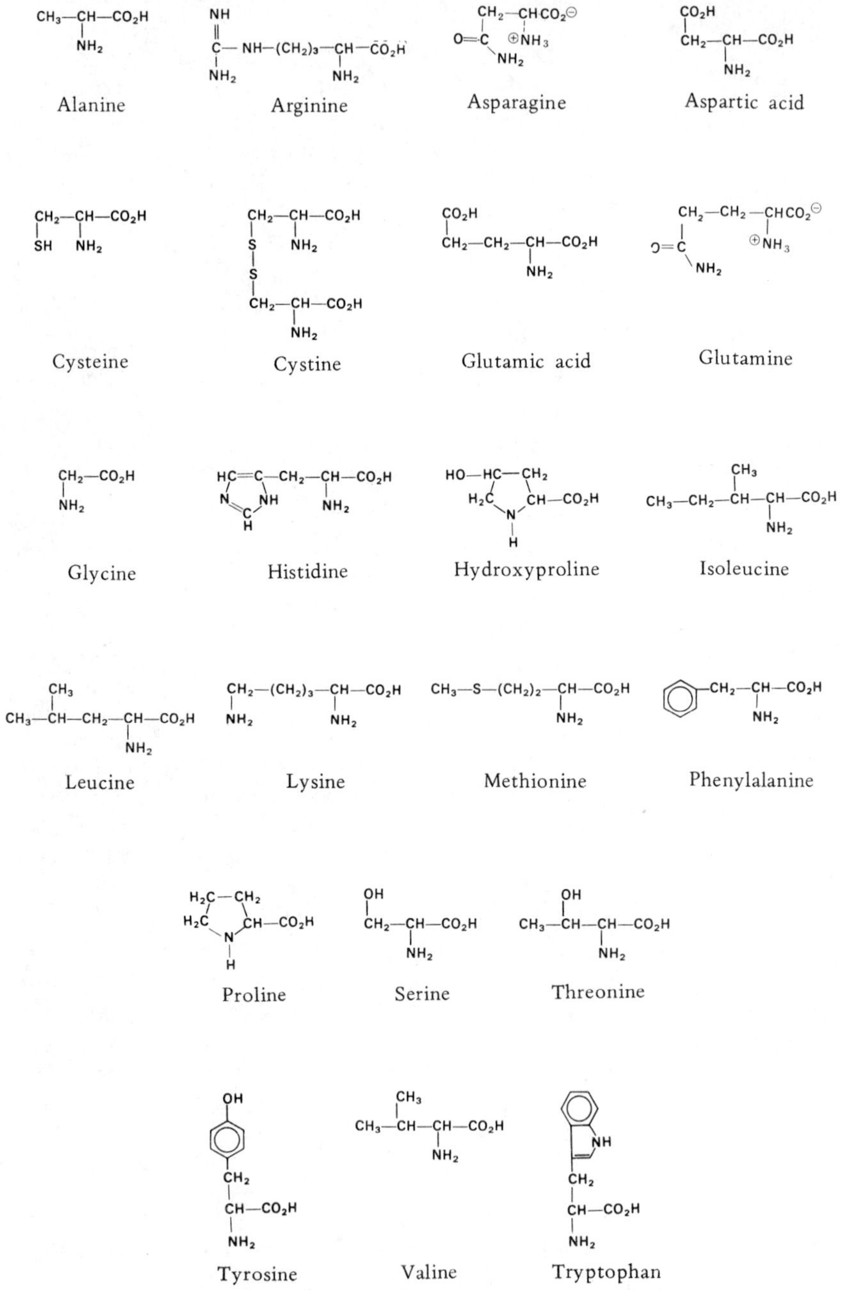

Figure 18.7. Common Amino Acids in Proteins.

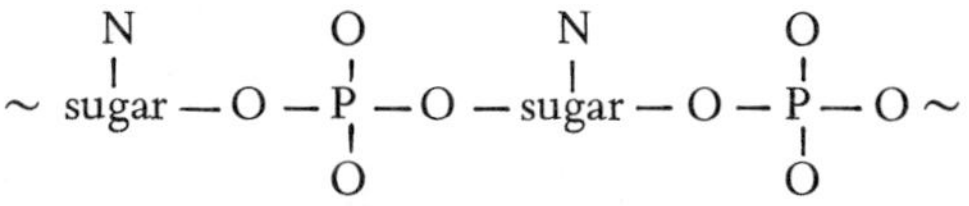

$$\sim \text{sugar} - O - \overset{\overset{\displaystyle N}{|}}{\underset{\underset{\displaystyle O}{|}}{P}} - O - \text{sugar} - O - \overset{\overset{\displaystyle N}{|}}{\underset{\underset{\displaystyle O}{|}}{P}} - O \sim$$

Figure 18.8. A Polynucleotide Chain.

The sugar in this chain is ribose or deoxyribose which will form two unique nucleic acids. One is called ribonucleic acid (RNA) and the other deoxyribonucleic acid (DNA). Figure 18.9 illustrates the structure of a DNA molecule.

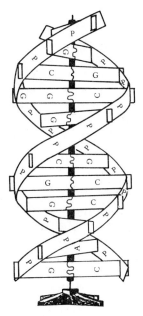

Figure 18.9. Structure of a DNA Molecule.

From our study of organic and biochemistry, it should be apparent that this is a massive area of chemistry and we have but scratched the surface. The understanding of the basic principles, however, is the key to future sophistication and advanced learning. The basic fundamentals have been given. It is your interest that will carry you from here.

Glossary

Amino Acid—A group of organic compounds, containing nitrogen, found in proteins. Amino Acids are produced by the hydrolysis of proteins.
DNA—An abbreviation for the substance deoxyribonucleic acid.

Enzyme—An organic substance produced by either a plant or an animal which causes chemical change in a substance by catalytic action.

Ester—A group of organic compounds formed from the reaction of a carboxylic acid with an alcohol. Fats and oils are examples of esters.

Glyceride—A special type fat or ester obtained from the alcohol glycerol.

Protein—A complex organic substance composed of a large number (10,000 M.W.) of amino acid units. Protein is found in all animal and vegetable matter.

RNA—An abbreviation for ribonucleic acid.

Exercises

1. What is the relationship between fats and esters?

2. Name the following esters:

    a. $C_2H_5CO_2C_3H_7$

    b. $CH_3CO_2C_2H_5$

    c. $CH_3CO_2CH_3$

    d. $C_3H_7CO_2C_5H_{11}$

3. Describe how amino acids relate to proteins.

4. Discuss the difference between a DNA molecule and an RNA molecule.

5. Predict the structure and name the product in the following reactions:

    a. $C_2H_5COOCH_3 \xrightarrow[\text{cat.}]{H_2}$ _____

    b. $C_3H_7COOC_2H_5 \xrightarrow{Na}_{C_2H_5OH}$ _____

    c. 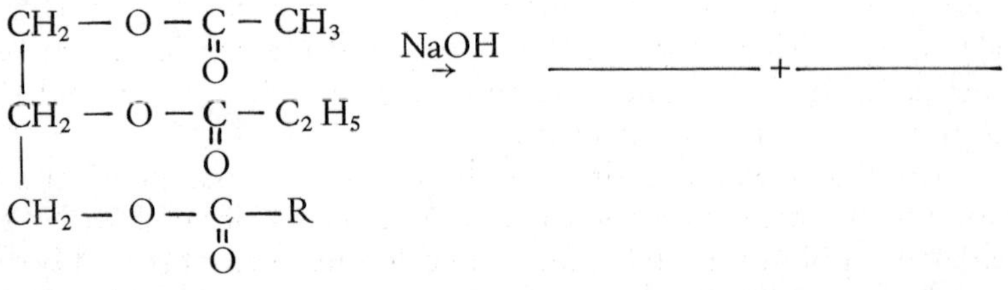
$$\begin{array}{l} CH_2 - O - \underset{\underset{O}{\|}}{C} - CH_3 \\ CH_2 - O - \underset{\underset{O}{\|}}{C} - C_2H_5 \\ CH_2 - O - \underset{\underset{O}{\|}}{C} - R \end{array} \xrightarrow{NaOH}$$
_____ + _____

6. Explain the difference between a soap and a synthetic detergent.

7. Draw the peptide linkage in a polypeptide chain and illustrate how the peptides and the amino acids are related to a given protein molecule.

8. What is the function of enzymes in an organism? What class of chemical compounds would the enzymes be listed under?

CHAPTER 19

Nuclear Chemistry

19.1 Introduction

One of the newest branches of chemistry, yet one of the most important in today's world, is nuclear chemistry. Everyone is aware of the atomic bomb, nuclear submarines and nuclear power stations. These are all the result of the understanding and control of nuclear chemistry.

Radioactivity, which is the emission of energy in the form of particles or rays by the disintegration of atomic nuclei, was first observed by Antoine Becquerel at the turn of the twentieth century. He, along with Roentgen, the Curies, and Rutherford contributed to the early discovery and understanding of radioactivity and thus the beginning of nuclear chemistry.

Roentgen was the first to observe x-rays, Bacquerel the first to observe the disintegration of uranium, the Curies discovered the two elements polonium and radium, and Rutherford observed both alpha and beta rays in the decay of uranium. The alpha and beta rays were first observed by Rutherford in 1899. The gamma ray was observed in 1900 by Paul Villard.

19.2 Alpha, Beta, and Gamma Rays

The *alpha* particle is a helium nucleus. That is, it consists of two protons and two neutrons. The charge is a positive two and the mass is about four atomic mass units (amu). There is a variety of symbols for the alpha particle. Some of the more common symbols are α, and $_2^4 He$.

The beta particle is an electron. The charge on a beta particle is negative. The beta particle is emitted when a neutron decomposes. Common symbols for the beta particle are β or $_{-1}^0 e$.

The *gamma* ray is a photon of energy which is much like an x-ray. Gamma rays do not have a charge and their mass is so small it is assumed to be negligible with respect to the other particles. The common symbol for the gamma ray is the Greek letter γ. Since gamma rays do not have mass or charge, the emission of a gamma ray does not alter the charge or the mass of an element. Table 19.1 summarizes the properties of alpha, beta, and gamma radiation.

TABLE 19.1
VARIOUS TYPES OF RADIATION

Type of Ray	Symbol	Charge	Mass
alpha	α	+2	4 amu
beta	β	−1	1/1837 amu
gamma	γ	0	negligible

19.3 Disintegration Series

Radioactive disintegration is the process by which an element loses an alpha or beta particle to create a new atomic species. As a basic example, let us consider some disintegrations involving only alpha particles. Remember that an alpha particle is equivalent to a helium nucleus.

$$_{92}^{238}U \rightarrow {}_{90}^{234}Th + {}_2^4He.$$

The above equation is the disintegration of uranium-238 to thorium-234. Notice that the atomic mass of uranium is reduced by four amu's and the atomic number is reduced from 92 to 90. Remember the atomic number is equal to the number of electrons or protons in the atom. Thus, if you were to add the alpha particle, $_2^4 He$, and $_{90}^{234}Th$ you would obtain $_{92}^{238}U$. Another example of this type would be $_{86}^{222}Rn$ going to $_{84}^{218}Po$. In equation form:

$$_{86}^{222}Rn \rightarrow {}_{84}^{218}Po + {}_2^4He.$$

Again, the mass is reduced by four amu's and the atomic number by two units of charge.

The other particle involved in a disintegration series is the beta particle $_{-1}^{0}e$. An example of this type of disintegration would be $_{83}^{210}$Bi going to $_{84}^{210}$Po. Remember that a beta particle is like an electron having negligible mass with respect to protons and a charge of negative one. Thus, the atomic mass remains constant and the atomic number is increased by one. In equation form:

$$_{83}^{210}Bi \rightarrow {}_{84}^{210}Po + {}_{-1}^{0}e.$$

Another example would be disintegration of $_{82}^{214}$Pb going to $_{83}^{214}$Bi. In equation form:

$$_{82}^{214}Pb \rightarrow {}_{83}^{214}Bi + {}_{-1}^{0}e.$$

If you were to study the periodic table, you would notice four naturally occurring radioactive decay series. All four series terminate with a nonradioactive form of lead. The first series is the Uranium Series. The second series is the Thorium Series, the third series is the Neptunium Series, and the fourth series is the Actinium Series. Each series proceeds from one atomic species to another by the systematic release of either an alpha particle or a beta particle. Let us now consider each of these series individually.

The Uranium Series begins with uranium-238 and terminates with lead. There are fifteen members in this series. The entire disintegration is shown in Figure 19.1. Notice the order of the release of alpha and beta particles. There is no pattern in the release. It is a series which must be followed step-by-step as there is more than one isotope of some of the series.

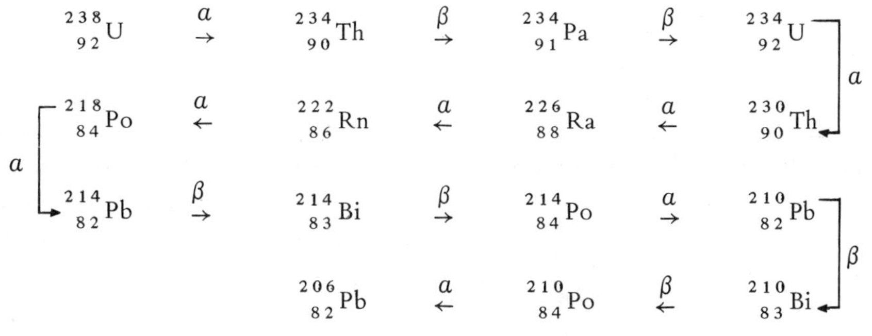

Figure 19.1. The Uranium Disintegration Series.

The second disintegration series is the Thorium Series. The Thorium Series begins with thorium-232 and terminates with lead-208. Figure 19.2 illustrates this series.

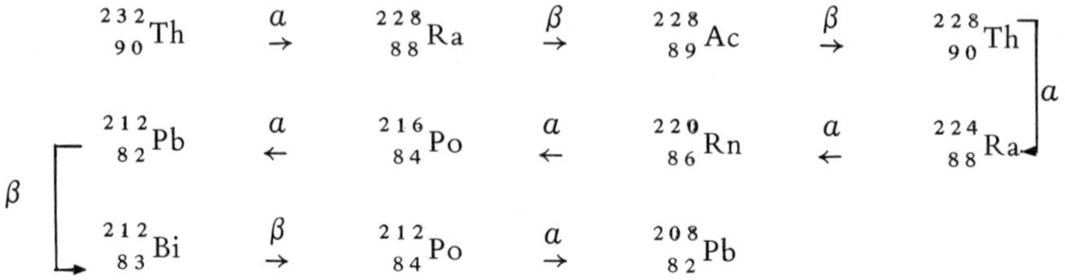

$${}_{90}^{232}\text{Th} \xrightarrow{\alpha} {}_{88}^{228}\text{Ra} \xrightarrow{\beta} {}_{89}^{228}\text{Ac} \xrightarrow{\beta} {}_{90}^{228}\text{Th} \bigg]\alpha$$

$$\beta \left[\; {}_{82}^{212}\text{Pb} \xleftarrow{\alpha} {}_{84}^{216}\text{Po} \xleftarrow{\alpha} {}_{86}^{220}\text{Rn} \xleftarrow{\alpha} {}_{88}^{224}\text{Ra} \right.$$

$${}_{83}^{212}\text{Bi} \xrightarrow{\beta} {}_{84}^{212}\text{Po} \xrightarrow{\alpha} {}_{82}^{208}\text{Pb}$$

Figure 19.2. The Thorium Disintegration Series.

The third disintegration series is the Actinium Series. The Actinium Series begins with uranium-235 and terminates with lead-207. Figure 19.3 illustrates the Actinium Series.

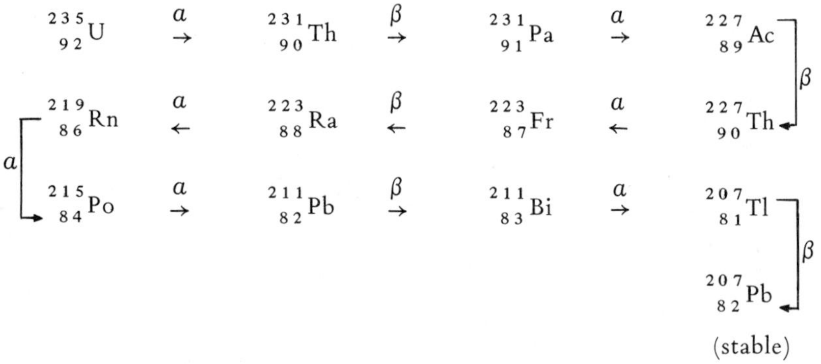

$${}_{92}^{235}\text{U} \xrightarrow{\alpha} {}_{90}^{231}\text{Th} \xrightarrow{\beta} {}_{91}^{231}\text{Pa} \xrightarrow{\alpha} {}_{89}^{227}\text{Ac} \bigg]\beta$$

$${}_{86}^{219}\text{Rn} \xleftarrow{\alpha} {}_{88}^{223}\text{Ra} \xleftarrow{\beta} {}_{87}^{223}\text{Fr} \xleftarrow{\alpha} {}_{90}^{227}\text{Th}$$

$$\alpha \left[\; {}_{84}^{215}\text{Po} \xrightarrow{\alpha} {}_{82}^{211}\text{Pb} \xrightarrow{\beta} {}_{83}^{211}\text{Bi} \xrightarrow{\alpha} {}_{81}^{207}\text{Tl} \right] \beta$$

$${}_{82}^{207}\text{Pb}$$

(stable)

Figure 19.3. The Actinium Disintegration Series.

The first three radioactive series that we have discussed are all naturally occurring. The fourth series called the Neptunium Series is an artificial series. That is, it has been produced by artificially induced radiation. The Neptunium Series begins with plutonium-241 and terminates with bismuth-209.

A term which is often used in discussing disintegration series is half-life. The *half-life* of a substance is the time required for one-half of a given amount of a radioactive element to disintegrate. The half-life of an element can vary from less than a second to billions of years. By knowing the half-life of an element, the radioactive dating of various substances can be done. Carbon-14 is often used to determine the age of a substance.

To illustrate the differences in half-life, study Table 19.2 which lists the half-lives for the elements in the three naturally occurring disintegration series.

TABLE 19.2
HALF-LIVES OF THE ELEMENTS IN THE
THREE NATURALLY OCCURRING SERIES

Uranium Series		Thorium Series		Actinium Series	
Element	Half-Life	Element	Half-Life	Element	Half-Life
$^{238}_{92}U$	4.50×10^9 yrs	$^{232}_{90}Th$	1.39×10^{10} yrs	$^{235}_{92}U$	7.10×10^8 yrs
$^{234}_{90}Th$	24.1 days	$^{228}_{88}Ra$	6.70 yrs	$^{231}_{90}Th$	24.6 hrs
$^{234}_{91}Pa$	1.18 min	$^{228}_{89}Ac$	6.13 hrs	$^{231}_{91}Pa$	3.43×10^4 yrs
$^{234}_{92}U$	2.50×10^5 yrs	$^{228}_{90}Th$	1.90 yrs	$^{227}_{89}Ac$	22.0 yrs
$^{230}_{90}Th$	8.00×10^4 yrs	$^{224}_{88}Ra$	3.64 days	$^{227}_{90}Th$	18.6 days
$^{226}_{88}Ra$	1620 yrs	$^{220}_{86}Rn$	54.5 sec	$^{223}_{87}Fr$	21.0 min
$^{222}_{86}Rn$	3.82 days	$^{216}_{84}Po$	0.16 sec	$^{223}_{88}Ra$	11.2 days
$^{218}_{84}Po$	3.05 min	$^{212}_{82}Pb$	10.6 hrs	$^{219}_{86}Rn$	3.92 sec
$^{214}_{82}Pb$	26.8 min	$^{212}_{83}Bi$	47.0 min	$^{215}_{84}Po$	1.83×10^{-3} sec
$^{214}_{83}Bi$	19.7 min	$^{212}_{84}Po$	3.0×10^{-7} sec	$^{211}_{82}Pb$	36.1 min
$^{214}_{84}Po$	1.64×10^{-4} sec	$^{208}_{82}Pb$	STABLE	$^{211}_{83}Bi$	2.16 min
$^{210}_{82}Pb$	22.0 yrs			$^{207}_{81}Tl$	4.79 min
$^{210}_{83}Bi$	5.00 days			$^{207}_{82}Pb$	STABLE
$^{210}_{84}Po$	138 days				
$^{206}_{82}Pb$	STABLE				

19.4 Artificial Radioactivity

The first artificially produced radioactive isotope was made by the daughter of Madame Curie in 1934. The isotope, $^{30}_{15}P$, was produced by the bombardment of $^{27}_{13}Al$ with alpha particles. In equation form:

$$^{27}_{13}\text{Al} + {}^{4}_{2}\text{He} \rightarrow {}^{30}_{15}\text{P} + {}^{1}_{0}n.$$
$$\text{(neutron)}$$

The charge and mass are balanced by the release of a neutron in the reaction only if they are indicated. It was a reaction similar to this in 1932 which had lead to the discovery of the neutron by an English scientist, Chadwick.

In general, the process of converting one element to another is called *transmutation*. Not all transmutations produce radioactive isotopes. For example, in 1919 Rutherford bombarded nitrogen atoms with alpha particles to produce an isotope of oxygen. The reaction in equation form was:

$$^{14}_{7}\text{N} + {}^{4}_{2}\text{He} \rightarrow {}^{17}_{8}\text{O} + {}^{1}_{1}\text{H} \,(\text{a proton})$$

However, many of the radioactive isotopes produced by transmutations are used in chemistry, geology, and medicine today.

Some of the important artificially produced isotopes are listed in Table 19.3.

TABLE 19.3
ISOTOPES PRODUCED BY TRANSMUTATION

Isotope	Half-Life
$^{14}_{6}\text{C}$	5570 years
$^{32}_{15}\text{P}$	14.3 days
$^{35}_{16}\text{S}$	87.0 days
$^{60}_{27}\text{Co}$	5.30 years
$^{90}_{38}\text{Sr}$	28.0 years
$^{31}_{53}\text{I}$	8.10 days
$^{137}_{87}\text{Cs}$	27.0 years

19.5 Nuclear Fission

When atomic energy is discussed in terms of use or application, the concept of nuclear fission is introduced. *Nuclear fission* is the process by which a heavy nucleus such as uranium-235 is bombarded by neutrons causing a reaction which splits the nucleus and leads to the production of new isotopes with a smaller mass. In stepwise fashion, the process of nuclear fission begins with the heavy nucleus accepting the neutron; then splitting into fragments whose mass may vary from 70 to 160 amu's. In the process, neutrons are also produced which in turn bombard different heavy nuclei producing a continuous reaction. This continuous type of reaction is called a *chain reaction*. A classic example of nuclear fission and a chain reaction is the bombardment of U-235 with neutrons. Many products are produced in the reaction. Some of these products are illustrated in Figure 19.4 which shows graphically the bombardment of U-235 by a neutron to produce nuclear fission and the chain reaction.

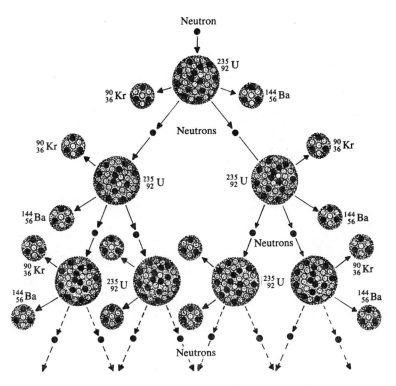

Figure 19.4. Nuclear Fission of Uranium-235.

19.6 Nuclear Fusion

As the name implies, the process of *nuclear fusion* is the chemical combination of two light weight nuclei to produce a heavier nucleus. Astronomers feel that fusion is the process by which energy is produced by stars. A great deal of energy is released in a fusion reaction.

A typical fusion reaction is the combining of the two isotopes of hydrogen called deuterium ($_1^2H$), and tritium ($_1^3H$). In equation form:

$$_1^2H + _1^3H \rightarrow _2^4He + _0^1n.$$

In both a fission and a fusion reaction, a neutron is produced. In order to have a fusion type reaction, an extremely high temperature is required. The temperature is so high that fusion reactions are triggered by a fission type reaction such as the neutron bombardment of U-235. Since the fission reaction is a combining reaction, there is no limit to the size of such a reaction.

19.7 Applications of Atomic Energy

Unfortunately, in today's world when atomic energy is mentioned, one immediately thinks of the atomic bomb or the hydrogen bomb. The atomic bomb is a fission reaction with the initial source being Uranium-235 and plutonium-239. Factors such as heat, radiation, and the explosion itself are so intense that this is one of the most destructive devices ever conceived by man. However, even worse is the hydrogen bomb for there are no limits to the size or potential of a hydrogen bomb which functions on the principle of fusion. Let us hope that common sense prevails.

The use of atomic energy in nuclear power plants to produce electricity is becoming more popular. As we become more concerned with our environment, the use of atomic energy as a source of fuel for these plants is tremendous as the only pollution we must be concerned with is the radioactive wastes. Atomic energy is the fuel of the future.

Another use of atomic energy or more specifically, radioisotopes, is in the carbon dating of the age of material. Since all living organisms contain both carbon-12 and carbon-14, a calculation of the ratio between these two isotopes can be used to determine the age of the organism. This is possible because when an organism dies, no additional carbon-14 is absorbed and that amount which is present in the organism will begin to decay according to its half-life of 5570 years. Thus, by setting a ratio of the carbon isotopes for the living and the dead organism, the age of the organism can be obtained.

Another form of atomic energy is still being used as a cure or preventive measure for some types of cancer. Cobalt or x-ray treatment has been found to help in the control but by no means is it a cure for cancer. The concept behind the use of cobalt treatments is that the radiation will kill the cancerous cell and thus allow the body to regenerate these cells with new, healthy material. The cobalt treatment will also kill any healthy cells which are also exposed. Thus, x-ray treatment is very limited.

More and more uses are being found for atomic energy. Let us hope man uses his ability in a constructive manner.

Glossary

Alpha Particle—An atomic particle consisting of two protons and two neutrons with a net charge of a positive two. It is a helium nucleus.

Beta Particle—An electron emitted from the nucleus of a radioactive atom.

Nuclear Fission—The process by which a heavy nucleus is split to form isotopes with a smaller mass.

Nuclear Fusion—The process by which two nuclei of light mass combine to form a nucleus of heavier mass.

Gamma Ray—A photon released from the nucleus. It is similar to an x-ray. There is no charge or mass associated with a gamma ray.

Half-Life—The time required for one-half of a given amount of a radioactive element to disintegrate.

Radioactivity—The emission of energy in the form of particles or rays by the disintegration or transmutation of atomic nuclei.

Transmutation—The process of changing one element to another by the bombardment of an atom with subatomic particles such as alpha and beta particles or neutrons.

Exercises

1. What is the difference between an artificial radioactive decay series and a natural radioactive decay series?

2. Why are nuclear reactions not discussed in the same terms as a regular chemical reaction? Explain in terms of conservation of mass and energy.

3. Explain the difference between nuclear fusion and nuclear fission. Give examples of each.

4. Explain why the mass of a beta particle can be assumed to be negligible.

5. A 40.0 gram sample of sulfur-35 has a half-life of 87.0 days. How many days will it require for the initial sample to contain 10.0 grams of sulfur-35?

6. In 1950 a student had a 15.0 gram sample of Cs-137. Assuming the same day in 1977, how many grams of Cs-137 are remaining? The half-life of Cs-137 is 27 years.

7. If thorium-232 undergoes decomposition, what is the final product?

8. Compare the three isotopes of hydrogen, $_1^1H$, $_1^2H$, and $_1^3H$ in terms of atomic number, atomic mass, number of protons, number of electrons, and the number of neutrons. Name the three isotopes.

9. Why is it necessary to initiate a nuclear fission reaction with a nuclear fusion reaction? Which reaction may be controlled?

10. Complete the following reactions:

a. $_{92}^{238}U \rightarrow _{90}^{234}Th + $ _____?_____

b. $_{82}^{210}Pb \rightarrow$ _____?_____ $+ _{-1}^{0}e$

c. $_8^{16}O + $ __?__ $_6^{13}C + _2^4He$

d. _____ $+ _2^4He \rightarrow _{15}^{30}P + _0^1n$

e. $_{11}^{23}Na + $ __?__ $\rightarrow _{12}^{23}Mg + _0^1n$

Appendixes

APPENDIX A

Alphabetical Listing of the Elements

[ ] indicates the most stable or best-known isotope

Element	Symbol	Atomic Number	Atomic Mass
Actinium	Ac	89	[227]
Aluminum	Al	13	26.98
Americium	Am	95	[243]
Antimony	Sb	51	121.75
Argon	Ar	18	39.95
Arsenic	As	33	74.92
Astatine	At	85	[210]
Barium	Ba	56	137.34
Berkelium	Bk	97	[249]
Beryllium	Be	4	9.01
Bismuth	Bi	83	208.98
Boron	B	5	10.81
Bromine	Br	35	79.91
Cadmiun	Cd	48	112.40
Calcium	Ca	20	40.08
Californium	Cf	98	[251]
Carbon	C	6	12.01
Cerium	Ce	58	140.12
Cesium	Cs	55	132.91
Chlorine	Cl	17	35.45
Chromium	Cr	24	52.00
Cobalt	Co	27	58.93
Copper	Cu	29	63.54
Curium	Cm	96	[247]
Dysprosium	Dy	66	162.50
Einsteinium	Es	99	[254]
Erbium	Er	68	167.26
Europium	Eu	63	151.96
Fermium	Fm	100	[253]
Fluorine	F	9	19.00
Francium	Fr	87	[223]
Gadolinium	Gd	64	157.25

Element	Symbol	Atomic Number	Atomic Mass
Gallium	Ga	31	69.72
Germanium	Ge	32	72.59
Gold	Au	79	196.97
Hafnium	Hf	72	178.49
Helium	He	2	4.00
Holmium	Ho	67	164.93
Hydrogen	H	1	1.01
Indium	In	49	114.82
Iodine	I	53	126.90
Iridium	Ir	77	192.20
Iron	Fe	26	55.85
Krypton	Kr	36	83.80
Lanthanum	La	57	138.91
Lawrencium	Lw	103	[257]
Lead	Pb	82	207.19
Lithium	Li	3	6.94
Lutetium	Lu	71	174.97
Magnesium	Mg	12	24.31
Manganese	Mn	25	54.94
Mendelevium	Md	101	[256]
Mercury	Hg	80	200.59
Molybdenum	Mo	42	95.94
Neodymium	Nd	60	144.24
Neon	Ne	10	20.18
Neptunium	Np	93	[237]
Nickel	Ni	28	58.71
Niobium	Nb	41	92.91
Nitrogen	N	7	14.01
Nobelium	No	102	[254]
Osmium	Os	76	190.20
Oxygen	O	8	16.00
Palladium	Pd	46	106.40
Phosphorus	P	15	30.97
Platinum	Pt	78	195.09
Plutonium	Pu	94	[242]
Polonium	Po	84	[210]
Potassium	K	19	39.10
Praseodymium	Pr	59	140.91
Promethium	Pm	61	[147]
Protactinium	Pa	91	[231]
Radium	Ra	88	[226]

Element	Symbol	Atomic Number	Atomic Mass
Radon	Rn	86	[222]
Rhenium	Re	75	186.20
Rhondium	Rh	45	102.91
Rubidium	Rb	37	85.47
Ruthenium	Ru	44	101.07
Samarium	Sm	62	150.35
Scandium	Sc	21	44.96
Selenium	Se	34	78.96
Silicon	Si	14	28.09
Silver	Ag	47	107.87
Sodium	Na	11	22.99
Strontium	Sr	38	87.62
Sulfur	S	16	32.06
Tantalum	Ta	73	180.95
Technetium	Tc	43	[99]
Tellurium	Te	52	127.60
Terbium	Tb	65	158.92
Thallium	Tl	81	204.37
Thorium	Th	90	232.04
Thulium	Tm	69	168.93
Tin	Sn	50	118.69
Titanium	Ti	22	47.90
Tungsten	W	74	183.85
Uranium	U	92	238.03
Vanadium	V	23	50.94
Xenon	Xe	54	131.30
Ytterbium	Yb	70	173.04
Yttrium	Y	39	88.91
Zinc	Zn	30	65.37
Zirconium	Zr	40	91.22

APPENDIX B

Alphabetical Listing of the Common Ions

Cation	Formula	Anion	Formula
Aluminum	Al^{3+}	Acetate	CH_3COO^-
Ammonium	NH_4^+	Borate	BO_3^{3-}
Barium	Ba^{2+}	Bromide	Br^-
Bismuth	Bi^{3+}	Carbonate	CO_3^{2-}
Cadmium	Cd^{2+}	Chlorate	ClO_3^-
Calcium	Ca^{2+}	Chloride	Cl^-
Chromium(II)	Cr^{2+}	Chlorite	ClO_2^-
Chromium(III)	Cr^{3+}	Chromate	CrO_4^{2-}
Cobalt(II)	Co^{2+}	Dichromate	$Cr_2O_7^{2-}$
Cobalt(III)	Co^{3+}	Dihydrogen phosphate	$H_2PO_4^-$
Copper(I)	Cu^+	Ferricyanide	$Fe(CN)_6^{3-}$
Copper(II)	Cu^{2+}	Ferrocyanide	$Fe(CN)_6^{4-}$
Iron(II)	Fe^{2+}	Fluoride	F^-
Iron(III)	Fe^{3+}	Bicarbonate	HCO_3^-
Lead(II)	Pb^{2+}	Bisulfate	HSO_4^-
Lithium	Li^+	Bisulfide	HS^-
Magnesium	Mg^{2+}	Bisulfite	HSO_3^-
Manganese(II)	Mn^{2+}	Hydrogen	H^+
Manganese(III)	Mn^{3+}	Hydroxide	OH^-
Mercury(I)	Hg_2^{2+}	Hypochlorite	ClO^-
Mercury(II)	Hg^{2+}	Iodide	I^-
Nickel(II)	Ni^{2+}	Metaphosphate	PO_3^-
Nickel(III)	Ni^{3+}	Biphosphate	HPO_4^{2-}
Potassium	K^+	Nitrate	NO_3^-
Scandium	Sc^{3+}	Nitrate	NO_2^-
Silver	Ag^+	Oxalate	$C_2O_4^{2-}$
Sodium	Na^+	Perchlorate	ClO_4^-
Strontium	Sr^{2+}	Peroxydisulfate	$S_2O_8^{2-}$
Tin(II)	Sn^{2+}	Phosphate	PO_4^{3-}
Tin(IV)	Sn^{4+}	Phosphite	HPO_3^{2-}
Zinc	Zn^{2+}	Sulfate	SO_4^{2-}
		Sulfide	S^{2-}
		Sulfite	SO_3^{2-}
		Thiosulfate	$S_2O_3^{2-}$

APPENDIX C

Equilibrium Constants (25°C)

Substance	Equation	K
Water	$H_2O = H^+ + OH^-$	1.00×10^{-14}
Acetic acid	$HC_2H_3O_2 = H^+ + C_2H_3O_2^-$	1.8×10^{-5}
Boric acid	$H_3BO_3 = H^+ + H_2BO_3^-$	6.0×10^{-10}
Carbonic ($CO_2 + H_2O$) acid	$H_2CO_3 = H^+ + HCO_3^-$	$K_1 : 4.4 \times 10^{-7}$
	$HCO_3^- = H^+ + CO_3^{--}$	$K_2 : 4.7 \times 10^{-11}$
Chromic acid	$H_2CrO_4 = H^+ + HCrO_4^-$	$K_1 : 2 \times 10^{-1}$
	$HCrO_4^- = H^+ + CrO_4^{--}$	$K_2 : 3.2 \times 10^{-7}$
Formic acid	$HCHO_2 = H^+ + CHO_2^-$	2.1×10^{-4}
Hydrocyanic acid	$HCN = H^+ + CN^-$	4×10^{-10}
Hydrofluoric acid	$HF = H^+ + F^-$	6.9×10^{-4}
Hydrogen peroxide	$H_2O_2 = H^+ + HO_2^-$	2.4×10^{-12}
Hydrogen sulfate ion	$HSO_4^- = H^+ + SO_4^{--}$	$K_2 : 1.2 \times 10^{-2}$
Hydrogen sulfide	$H_2S = H^+ + HS^-$	$K_1 : 1.0 \times 10^{-7}$
	$HS^- = H^+ + S^{--}$	$K_2 : 1.3 \times 10^{-13}$
Nitrous acid	$HNO_2 = H^+ + NO_2^-$	4.5×10^{-4}
Oxalic acid	$H_2C_2O_4 = H^+ + HC_2O_4^-$	$K_1 : 3.8 \times 10^{-2}$
	$HC_2O_4^- = H^+ + C_2O_4^{--}$	$K_2 : 5.0 \times 10^{-5}$
Phosphoric acid	$H_3PO_4 = H^+ + H_2PO_4^-$	$K_1 : 7.1 \times 10^{-3}$
	$H_2PO_4^- = H^+ + HPO_4^{--}$	$K_2 : 6.3 \times 10^{-8}$
	$HPO_4^{--} = H^+ + PO_4^{---}$	$K_3 : 4.4 \times 10^{-13}$
Phosphorous acid	$H_2HPO_3 = H^+ + HHPO_3^-$	$K_1 : 1.6 \times 10^{-2}$
Sulfurous ($SO_2 + H_2O$) acid	$H_2SO_3 = H^+ + HSO_3^-$	$K_1 : 1.2 \times 10^{-2}$
	$HSO_3^- = H^+ + SO_3^{--}$	$K_2 : 5.6 \times 10^{-8}$
Zinc ion	$Zn(H_2O)_4^{++} = H^+ + Zn(H_2O)_3OH^+$	2.5×10^{-10}
Ammonium hydroxide	$NH_4OH = NH_4^+ + OH^-$	1.8×10^{-5}
Barium hydroxide	$Ba(OH)_2 = BaOH^+ + OH^-$	strong
	$BaOH^+ = Ba^{++} + OH^-$	$K_2 : 1.4 \times 10^{-1}$
Calcium hydroxide	$Ca(OH)_2 = CaOH^+ + OH^-$	strong
	$CaOH^+ = Ca^{++} + OH^-$	$K_2 : 3.5 \times 10^{-2}$

APPENDIX D

Solubility Product Constants (20–25°C)

Substance	K_{sp}	Substance	K_{sp}
$AgCH_3COO$	4.0×10^{-3}	FeS	1.0×10^{-17}
Ag_2CO_3	8.0×10^{-12}	Hg_2Cl_2	1.1×10^{-18}
AgCN	1.0×10^{-16}	Hg_2SO_4	6.0×10^{-7}
AgCl	1.8×10^{-10}	HgS	1.0×10^{-50}
Ag_2CrO_4	2.0×10^{-12}	MgF_2	8.0×10^{-8}
AgBr	5.0×10^{-13}	$MgCO_3$	4.0×10^{-5}
AgI	8.5×10^{-17}	$Mg(OH)_2$	9.0×10^{-12}
Ag_2S	1.0×10^{-50}	MgC_2O_4	8.6×10^{-5}
Ag_2SO_4	1.7×10^{-5}	$MnCO_3$	9.0×10^{-11}
$Al(OH)_3$	1.0×10^{-33}	$Mn(OH)_2$	2.0×10^{-13}
$BaCO_3$	1.6×10^{-9}	MnS	1.0×10^{-13}
$BaCrO_4$	1.2×10^{-10}	NiS	1.0×10^{-22}
BaC_2O_4	1.5×10^{-8}	$PbCl_2$	1.6×10^{-5}
$BaSO_4$	1.5×10^{-9}	PbI_2	8.3×10^{-9}
$CaCO_3$	4.8×10^{-9}	$PbCO_3$	1.5×10^{-13}
$Ca(OH)_2$	1.3×10^{-6}	$PbCrO_4$	3.6×10^{-5}
CaC_2O_4	1.3×10^{-9}	$Pb(OH)_2$	4.0×10^{-15}
$CaSO_4$	2.4×10^{-5}	$PbSO_4$	1.3×10^{-8}
CdS	1.0×10^{-26}	PbS	1.0×10^{-26}
CoS	1.0×10^{-21}	$Sn(OH)_2$	1.0×10^{-27}
$Cr(OH)_3$	1.0×10^{-30}	SnS	1.0×10^{-27}
$CuCO_3$	2.5×10^{-10}	$SrCO_3$	7.0×10^{-10}
CuCl	3.2×10^{-7}	$SrCrO_4$	3.6×10^{-5}
$Cu(OH)_2$	2.0×10^{-19}	SrC_2O_4	5.6×10^{-8}
CuS	1.0×10^{-36}	$SrSO_4$	7.6×10^{-7}
$FeCO_3$	2.0×10^{-11}	$Zn(OH)_2$	5.0×10^{-17}
$Fe(OH)_2$	2.0×10^{-15}	ZnS	1.0×10^{-20}
$Fe(OH)_3$	1.0×10^{-37}		

APPENDIX E

Vapor Pressure of Water
at Various Temperatures

Temperature ($^\circ$C)	Vapor Pressure (*torr*)	Temperature ($^\circ$C)	Vapor Pressure (*torr*)
0	4.6	29	30.0
5	6.5	30	31.8
10	9.2	31	33.7
11	9.8	32	35.7
12	10.5	33	37.7
13	11.2	34	39.9
14	12.0	35	42.2
15	12.8	40	55.3
16	13.6	45	71.9
17	14.5	50	92.5
18	15.5	55	118.0
19	16.5	60	149.4
20	17.5	65	187.5
21	18.6	70	233.7
22	19.8	75	289.1
23	21.1	80	355.1
24	22.4	85	433.6
25	23.8	90	525.8
26	25.2	95	633.9
27	26.7	100	760.0
28	28.4		

APPENDIX F

Common Conversion Units

Mass and Weight:

1 kilogram = 1000 grams
1 gram = 1000 milligrams
1 pound = 454 grams
1 kilogram = 2.20 pounds
1 gram = 15.4 grains
1 ounce(av) = 28.3 grams
1 ounce(ap) = 31.1 grams

Length:

1 meter = 100 centimeters
1 centimeter = 10 millimeters
1 inch = 2.54 centimeters
1 meter = 39.4 inches
1 kilometer = 0.62 miles
1 Ångstrom = 1×10^{-8} cm
1 micron = 1×10^{-3} mm

Volume and Capacity:

1 liter = 1000 milliliters
1000 milliliters = 1000.028 cc
1 quart = 946 milliliters
1 liter = 1.06 quarts
1 ounce(US) = 29.6 milliliters
1 cubic foot = 28.3 liters

Pressure:

1 atmosphere = 760 torr
1 atmosphere = 29.92 inches of Hg
1 atmosphere = 14.70 lb per sq in

Temperature:

Absolute zero $(0°K) = -273.15°C$
$0°C = 32°F$
$0°C = 273°K$
$°K = °C + 273$
$°F = 9/5°C + 32$
$°C = 5/9(°F - 32)$

Electrical:

1 ampere = a flow of one coulomb per second
1 volt = the potential difference necessary
    to cause a current of one ampere
      through a resistance of one ohm
Charge of one electron = 1.6×10^{-19} coul

Miscellaneous:

Avogadro's number = $6.0x \times 10^{23}$
Gas Constant (R) = 0.082 1 atm per ° per mole
             = 1.99 cal per ° per mole
1 BTU = 252 calories
1 kilocalorie = 1000 calories

APPENDIX G

Four Place Logarithms

N	0	1	2	3	4	5	6	7	8	9
10	0000	0043	0086	0128	0170	0212	0253	0294	0334	0374
11	0414	0453	0492	0531	0569	0607	0645	0682	0719	0755
12	0792	0828	0864	0899	0934	0969	1004	1038	1072	1106
13	1139	1173	1206	1239	1271	1303	1335	1367	1399	1430
14	1461	1492	1523	1553	1584	1614	1644	1673	1703	1732
15	1761	1790	1818	1847	1875	1903	1931	1959	1987	2014
16	2041	2068	2095	2122	2148	2175	2201	2227	2253	2279
17	2304	2330	2355	2380	2405	2430	2455	2480	2504	2529
18	2553	2577	2601	2625	2648	2672	2695	2718	2742	2765
19	2788	2810	2833	2856	2878	2900	2923	2945	2967	2989
20	3010	3032	3054	3075	3096	3118	3139	3160	3181	3201
21	3222	3243	3263	3284	3304	3324	3345	3365	3385	3404
22	3424	3444	3464	3483	3502	3522	3541	3560	3579	3598
23	3617	3636	3655	3674	3692	3711	3729	3747	3766	3784
24	3802	3820	3838	3856	3874	3892	3909	3927	3945	3962
25	3979	3997	4014	4031	4048	4065	4082	4099	4116	4133
26	4150	4166	4183	4200	4216	4232	4249	4265	4281	4298
27	4314	4330	4346	4362	4378	4393	4409	4425	4440	4456
28	4472	4487	4502	4518	4533	4548	4564	4579	4594	4609
29	4624	4639	4654	4669	4683	4698	4713	4728	4742	4757
30	4771	4786	4800	4814	4829	4843	4857	4871	4886	4900
31	4914	4928	4942	4955	4969	4983	4997	5011	5024	5038
32	5051	5065	5079	5092	5105	5119	5132	5145	5159	5172
33	5185	5198	5211	5224	5237	5250	5263	5276	5289	5302
34	5315	5328	5340	5353	5366	5378	5391	5403	5416	5428
35	5441	5453	5465	5478	5490	5502	5514	5527	5539	5551
36	5563	5575	5587	5599	5611	5623	5635	5647	5658	5670
37	5682	5694	5705	5717	5729	5740	5752	5763	5775	5786
38	5798	5809	5821	5832	5843	5855	5866	5877	5888	5899
39	5911	5922	5933	5944	5955	5966	5977	5988	5999	6010
40	6021	6031	6042	6053	6064	6075	6085	6096	6107	6117
41	6128	6138	6149	6160	6170	6180	6191	6201	6212	6222
42	6232	6243	6253	6263	6274	6284	6294	6304	6314	6325
43	6335	6345	6355	6365	6375	6385	6395	6405	6415	6425
44	6435	6444	6454	6464	6474	6484	6493	6503	6513	6522
45	6532	6542	6551	6561	6571	6580	6590	6599	6609	6618
46	6628	6637	6646	6656	6665	6675	6684	6693	6702	6712
47	6721	6730	6739	6749	6758	6767	6776	6785	6794	6803
48	6812	6821	6830	6839	6848	6857	6866	6875	6884	6893
49	6902	6911	6920	6928	6937	6946	6955	6964	6972	6981
50	6990	6998	7007	7016	7024	7033	7042	7050	7059	7067
51	7076	7084	7093	7101	7110	7118	7126	7135	7143	7152
52	7160	7168	7177	7185	7193	7202	7210	7218	7226	7235
53	7243	7251	7259	7267	7275	7284	7292	7300	7308	7316
54	7324	7332	7340	7348	7356	7364	7372	7380	7388	7396

Four Place Logarithms (Cont.)

N	0	1	2	3	4	5	6	7	8	9
55	7404	7412	7419	7427	7435	7443	7451	7459	7466	7474
56	7482	7490	7497	7505	7513	7520	7528	7536	7543	7551
57	7559	7566	7574	7582	7589	7597	7604	7612	7619	7627
58	7634	7642	7649	7657	7664	7672	7679	7686	7694	7701
59	7709	7716	7723	7731	7738	7745	7752	7760	7767	7774
60	7782	7789	7796	7803	7810	7818	7825	7832	7839	7846
61	7853	7860	7868	7875	7882	7889	7896	7903	7910	7917
62	7924	7931	7938	7945	7952	7959	7966	7973	7980	7987
63	7993	8000	8007	8014	8021	8028	8035	8041	8048	8055
64	8062	8069	8075	8082	8089	8096	8102	8109	8116	8122
65	8129	8136	8142	8149	8156	8162	8169	8176	8182	8189
66	8195	8202	8209	8215	8222	8228	8235	8241	8248	8254
67	8261	8267	8274	8280	8287	8293	8299	8306	8312	8319
68	8325	8331	8338	8344	8351	8357	8363	8370	8376	8382
69	8388	8395	8401	8407	8414	8420	8426	8432	8439	8445
70	8451	8457	8463	8470	8476	8482	8488	8494	8500	8506
71	8513	8519	8525	8531	8537	8543	8549	8555	8561	8567
72	8573	8579	8585	8591	8597	8603	8609	8615	8621	8627
73	8633	8639	8645	8651	8657	8663	8669	8675	8681	8686
74	8692	8698	8704	8710	8716	8722	8727	8733	8739	8745
75	8751	8756	8762	8768	8774	8779	8785	8791	8797	8802
76	8808	8814	8820	8825	8831	8837	8842	8848	8854	8859
77	8865	8871	8876	8882	8887	8893	8899	8904	8910	8915
78	8921	8927	8932	8938	8943	8949	8954	8960	8965	8971
79	8976	8982	8987	8993	8998	9004	9009	9015	9020	9025
80	9031	9036	9042	9047	9053	9058	9063	9069	9074	9079
81	9085	9090	9096	9101	9106	9112	9117	9122	9128	9133
82	9138	9143	9149	9154	9159	9165	9170	9175	9180	9186
83	9191	9196	9201	9206	9212	9217	9222	9227	9232	9238
84	9243	9248	9253	9258	9263	9269	9274	9279	9284	9289
85	9294	9299	9304	9309	9315	9320	9325	9330	9335	9340
86	9345	9350	9355	9360	9365	9370	9375	9380	9385	9390
87	9395	9400	9405	9410	9415	9420	9425	9430	9435	9440
88	9445	9450	9455	9460	9465	9469	9474	9479	9484	9489
89	9494	9499	9504	9509	9513	9518	9523	9528	9533	9538
90	9542	9547	9552	9557	9562	9566	9571	9576	9581	9586
91	9590	9595	9600	9605	9609	9614	9619	9624	9628	9633
92	9638	9643	9647	9652	9657	9661	9666	9671	9675	9680
93	9685	9689	9694	9699	9703	9708	9713	9717	9722	9727
94	9731	9736	9741	9745	9750	9754	9759	9763	9768	9773
95	9777	9782	9786	9791	9795	9800	9805	9809	9814	9818
96	9823	9827	9832	9836	9841	9845	9850	9854	9859	9863
97	9868	9872	9877	9881	9886	9890	9894	9899	9903	9908
98	9912	9917	9921	9926	9930	9934	9939	9943	9948	9952
99	9956	9961	9965	9969	9974	9978	9983	9987	9991	9996

Standard Electrode Potentials at 25°C

Half-Reaction	Potential (volts)
$Li^+ + e^- = Li$	-3.04
$Na^+ + e^- = Na$	-2.71
$Mg^{2+} + 2e^- = Mg$	-2.37
$Al^{3+} + 3e^- = Al$	-1.66
$Mn^{2+} + 2e^- = Mn$	-1.18
$Zn^{2+} + 2e^- = Zn$	-0.76
$Cr^{3+} + 3e^- = Cr$	-0.74
$Fe^{2+} + 2e^- = Fe$	-0.44
$Cd^{2+} + 2e^- = Cd$	-0.40
$Co^{2+} + 2e^- = Co$	-0.28
$Ni^{2+} + 2e^- = Ni$	-0.25
$Sn^{2+} + 2e^- = Sn$	-0.14
$Pb^{2+} + 2e^- = Pb$	-0.13
$2H^+ + 2e^- = H_2$	0.00 (Definition)
$Cu^{2+} + 2e^- = Cu$	0.34
$I_2 + 2e^- = 2I^-$	0.54
$Fe^{3+} + e^- = Fe^{2+}$	0.77
$Hg_2^{2+} + 2e^- = 2Hg$	0.79
$Ag^+ + e^- = Ag$	0.80
$2Hg^{2+} + 2e^- = Hg_2^{2+}$	0.92
$Cr_2O_7{}^{2-} + 14H^+ + 6e^- = 2Cr^{3+} + 7H_2O$	1.33
$Cl_2 + 2e^- = 2Cl^-$	1.36
$MnO_4 + 8H^+ + 5e^- = Mn^{2+} + 4H_2O$	1.51
$F_2 + 2e^- = 2F^-$	2.65

APPENDIX I

Answers to Odd-Numbered Exercises

Chapter 1

1. Man is a quantity of matter independent of location. Weight is a measure of matter dependent upon the force of gravity and varies with respect to location.

3. Physical changes in matter—The melting of ice, the dissolving of a salt in water, the crystallization of sugar, etc.

5. a. 3   d. 1
   b. 1   e. 5
   c. 6

7. a. 7.1×10^6 d. 7×10^{-6}
   b. 1.5×10^{36} e. 8.9×10^{-14}
   c. 7×10^6

9.

mm	cm	m	km
2240	224	2.24	2.24×10^{-3}
1.5×10^6	1.5×10^5	1.5×10^3	1.5
4.55×10^5	4.55×10^4	455	.455
14560	1456	14.56	.01456
9.6×10^8	9.6×10^7	9.6×10^5	9.6×10^2

11.

ml	l	cc
2.55×10^3	2.55	2.55×10^3
1940	1.940	1940
3550	3.550	3550
5.2×10^5	5.2×10^2	5.2×10^5

13.

°C	°K	°F
25	298	77
150	425	302
27	300	77
20	293	68
-261	17	-436

15. 32.004, 130.0, 5280, 25, 5,000

17. $g \left(\dfrac{cm}{sec} \right)^2$

19. 234 cc or 234 ml

Chapter 2

1. For clarity of symbols—to distinguish between molecules and atoms.
   Example: NO is nitric oxide and No is nobelium.

3. a. 36.5 amu/molecule
   b. 42 amu/molecule
   c. 81 amu/molecule
   d. 278 amu/molecule
   e. 74 amu/molecule

5. a. 71% Ca, 29% O
   b. 94% O, 6% H
   c. 66% Cu, 34% S
   d. 80% Cu, 2% S
   e. 27% Na, 16% N, 58% O

7. a. CH_3O
   b. CH_2O
   c. C_2H_6O
   d. FeS
   e. Fe_2O_3

9. a. 80% Cu, 20% S
   b. Cu_2S
   c. Cu_2S

Chapter 3

1. Alpha particles have a positive charge and were repelled by the nucleus, thus the nucleus had to also have a positive charge according to the rule that like charges repel each other.

3. F, Cl, Br, I

5. F_2, O_2, I_2

7. Astatine is a member of a radioactive decay series and as a result, has different properties than the alto halogens.

9. All members of the Lanthanide Series are radioactive. Atomic #58–#71.

Chapter 4

1. Refer to section 4.2.

3. A three-dimensional probability distribution for locating an electron about the nucleus of an atom.

5. Nitrogen

7. a. 2
   b. 2
   c. 8
   d. 8

9.

Element	At. #	Elec. Conf.
a. N	7	$1s^2\ 2s^2\ 2p^3$
b. B	5	$1s^2\ 2s^2\ 2p^1$
c. Mg	12	$1s^2\ 2s^2\ 2p^6\ 3s^2$
d. Li	3	$1s^2\ 2s^1$
e. C	6	$1s^2\ 2s^2\ 2p^2$

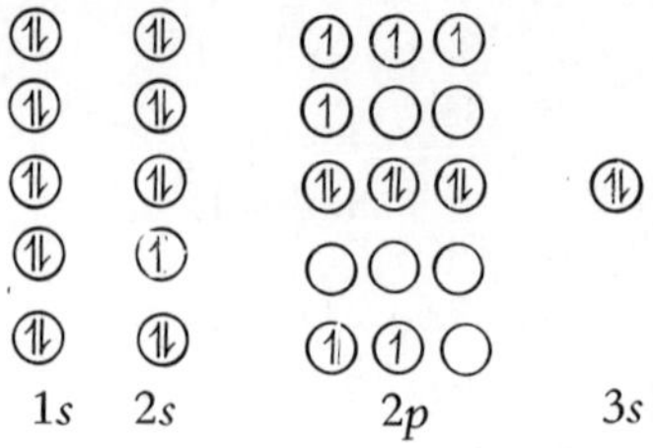

$1s \quad 2s \qquad 2p \qquad 3s$

11. See Glossary Chapter 4.

Chapter 5

1. a. F_2 $2p$

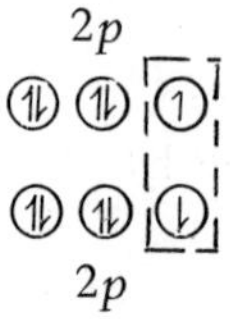

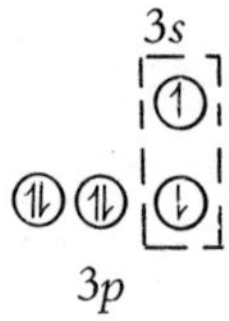

d. N = (Hybrid)

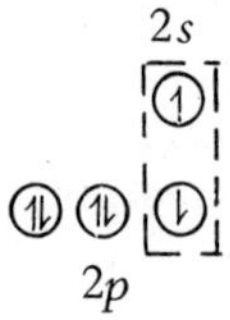

2 Oxygen $2p$ $2p$

b. LiF   Li = $2s$

F = $2p$

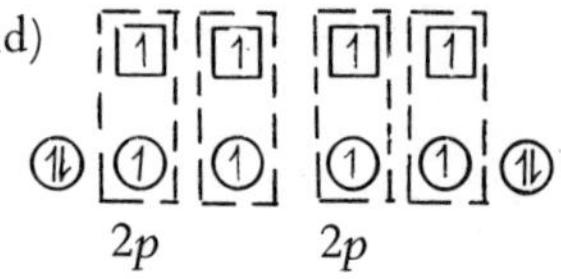

e. H $1s$

Cl $3p$

Na $3s$

c. Cl $3p$

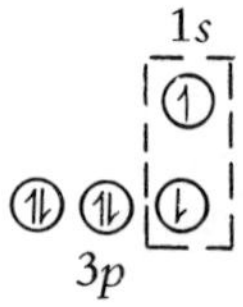

3. Increases

5. a. Bent
   b. Linear
   c. Tetrahedral
   d. Tetrahedral
   e. Linear

7. a. Chloride—anion
   b. Sodium—cation
   c. Acetate—anion
   d. Sulfate—anion
   e. Ammonium—cation

9. a. $Li^+\ F^-$
   b. $Co^{2+}\ Cl^-$
   c. $Hg_2^{\ +}$
   d. $Sn^{4+}\ Cl^-$
   e. $Br_2^{\ 0}$
   f. $Ag^0 = _+$
   g. $Ca^{2+}(OH)_2$
   h. $Li^+\ H^-$
   i. $Si^{4+}\ O_2^{\ 2-}$
   j. $Al^{3+}\ PO_4^{\ 3-}$

Chapter 6

1. a. Base
    b. Salt
    c. Acid

    d. Salt
    e. Salt

3. a. ClO_3^-
    b. CH_3COO^-
    c. SO_3^{2-}
    d. F^-
    e. IO_3^-

    f. BO_3^{3-}
    g. CO_3^{2-}
    h. $C_2O_4^{2-}$
    i. CN^-
    j. H^-

5. a. Fe^{2+}
    b. Pb^{2+}
    c. Cu^{2+}
    d. Hg^{2+}
    e. Hg_2^{2+}

    f. Co^{2+}
    g. Ni^{2+}
    h. Cu^+
    i. Mn^{2+}
    j. Mn^{4+}

7. a. HNO_2
    b. $HClO_4$
    c. HF

    d. HClO
    e. HI

9. a. Potassium hypochlorite
    b. Magnesium nitride
    c. Bismuth (III) chloride
    d. Calcium sulfite
    e. Potassium bisulfate

    f. Ammonium iodide
    g. Calcium acetate
    h. Copper (II) oxalate
    i. Alumenium carbonate
    j. Manganese (IV) sulfite

Chapter 7

1. a. $Ca CO_3 \rightarrow CaO + CO_2$
    b. $2H_2 + O_2 \rightarrow 2H_2O$
    c. $Zn + 2HCl \rightarrow H_2 + Zn Cl_2$
    d. $CH_3COOH + NH_4OH \rightarrow NH_4 CH_3 COO + H_2O$
    e. $H_2 + Br_2 \rightarrow 2HBr$
    f. $2HgO \rightarrow 2Hg + O_2$
    g. $2 Al_2O_3 \rightarrow 4 Al + 3 O_2$
    h. $2 P + 3 Cl_2 \rightarrow 2 PCl_3$

3. a. $2 H_2 + O_2 \rightarrow 2 H_2O$
    b. $Ca CO_3 \rightarrow Ca O + CO_2$
    c. $2 Al + 6 HCl \rightarrow 2AlCl_3 + 3 H_2$
    d. $Ca(OH)_2 + 2HCl \rightarrow CaCl_2 + 2 H_2O$

5. a. $N_2 + 3\,H_2 = 2\,NH_3$
    b. $CO + NO_2 = CO_2 + NO$
    c. $Pb\,I_{2\,(s)} = Pb^{2+}(aq) + 2\,I^-(aq)$
    d. $2\,C_2H_6 + 7\,O_2 = 6\,H_2O + 4\,CO_2$
    e. $2SO_2 + O_2 = 2\,SO_3$
    f. $4\,HCl + O_2 = 2\,H_2O + 2\,Cl_2$
    g. $C_{12}H_{22}O_{11} + 12\,O_2 = 12\,CO_2 + 11\,H_2O$
    h. $2\,HBr + Ca(OH)_2 = CaBr_2 + 2\,H_2O$
    i. $2\,NO + O_2 = 2\,NO_2$
    j. $2\,Rb + Br_2 = 2\,RbBr$
    k. $2\,Na + 2\,H_2O = 2\,NaOH + H_2$
    l. $Cl_2 + H_2O = HOCl + HCl$
    m. $2\,KNO_3 = 2\,KNO_2 + O_2$
    n. $Mg(NO_3)_2 + 2\,H_2SO_4 = 2\,HNO_3 + Mg(HSO_4)_2$
    o. $3\,CH_3\,COOH + Al(OH)_3 = Al(CH_3COO)_3 + 3\,H_2O$

Chapter 8

1. a. 24.3 grams   c. 26 grams
    b. 65.4 grams   d. 2×10^{-23} grams

3. a. 0.1 g–A   c. 0.2 g–A
    b. 17.8 g–A   d. 7.4 g–A

5. 1 Atom–1.7×10^{-24} gram-atoms

7. a. 1.2×10^{24} molecules   c. 7.2×10^{22} molecules
    b. 6×10^{22} molecules   d. 1.2×10^{24} molecules

9.

# g–A	# Atoms	# g
0.1	6×10^{22}	5.87
1.7×10^{-22}	100	1.0×10^{-20}
2.5	1.5×10^{24}	147

11. A, D, B, C

13. a. .2   c. .11
    b. .1   d. 6.6×10^{22} ions

15. a. 5 moles
    b. 3.0×10^{24}
    c. 6.6×10^{24}

17. $Cu\,SO_4 \cdot 5\,H_2O$

19. CuO

21. 10 moles H_2O

23. a. 2 moles
     b. 8.0 grams
25. 4.9 Kg, 2.7 liters

Chapter 9

1. 0.44 g/liter
3. 194 liters
5. 734.4 tons
7. .92 liters
9. 94 liters
11. 67 g/mole
13. 3.5 liters
15. .64 moles
17. 9.1 liters
19. 3 moles

Chapter 10

1. Solid—rigid, definite shape
    Liquid—shape of container, flows
    Gas—greatest molecular motion, no definite shape
3. Some solids are classified as metals and others as nonmetals. Nonmetals do not conduct electricity.
5. Six
7. 637 calories
9. Solid and liquid—0°C
11. 112.5 grams ICl
13. 19 kcal
15. 23°C

Chapter 11

1. M = .17 moles/kg
3. a. 2 M c. 0.5 M
    b. 1.28 M

5.   a. .025 M c. .067 M
     b. .0014 M

7.   a. 12 g   c. 9 g
     b. 33g

9.   a. Take 19 ml of 2 M HCl and add enough water to make 150 ml of solution.
     b. Take 125 ml of 6 M H_3PO_4 and add enough water to make 500 ml of 1.5 M H_3PO_4.
     c. Take 750 ml of 12 M NaOH and add enough water to make 1.5 liters of solution.

11. 1.5 liters

Chapter 12

1.   a. 0.1 mole   c. .03 mole
     b. 0.4 mole   d. = 10 moles

3.   3 moles

5.   40 g

7.   30 g

9.   a. 5.85 g
     b. 1.12 liters

11. 16.8 liters

13. 6.7 liters

15. .13 M

17. 26 grams

19. 2.2×10^{-5} N

Chapter 13

1.   a. $K_p = \dfrac{p(HI)^2}{(pH_2)\,(pI_2)}$ $K = \dfrac{[HI]^2}{[H_2]\,[I_2]}$

     b. $K_p = \dfrac{(p\,HBr)^4\,(p\,O_2)}{(p\,Br_2)^2\,(p\,H_2O)^2}$ $K = \dfrac{[HBr]^4\,[O_2]}{[Br_2]^2\,[H_2O]^2}$

3.   a. $K = \dfrac{[HCl]^2}{[H_2]\,[Cl_2]}$

     b. $K = \dfrac{[Cu^{2+}]\,[NO_3^{-}]}{[Ag^{+}]^2\,[NO_3^{-}]}$

c. $K = \dfrac{[H^+]\,[HSO_4{}^-]}{[H_2SO_4]}$

d. $K = \dfrac{[NH_4{}^+]\,[OH^-]}{[NH_4OH]}$

e. $K = [Al^{3+}]^2\,[SO_4{}^{2-}]^3$

5. a. Acidic   d. Acidic
   b. Neutral   e. Neutral
   c. Basic

7. $K = .58$

9. $4.0 \times 10^{-28} = K_{sp}$

Chapter 14

1. Reduced$-O + 2e^- = O^{2-}$

3. Li, Al, Cr, H_2, F^-

5. Any neutral species above Cl^- on an electrode potential chart. (Example: Ag, Cu, Pb, etc.)

7.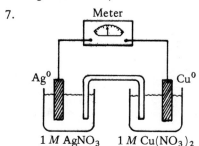

9. a. Fe^{2+} would be reduced and Mg^0 would be oxidized.
   b. No change would occur.

Chapter 15

1. a. Alcohol   d. Alkene
   b. Carboxylic acid   e. Aldehyde
   c. Ketone

3. a.
$$H-\overset{\displaystyle H}{\underset{\displaystyle H}{C}}-\overset{\displaystyle H}{\underset{\displaystyle H}{C}}-\overset{\displaystyle H}{\underset{\displaystyle H}{C}}-\overset{\displaystyle H}{\underset{\displaystyle H}{C}}-\overset{\displaystyle H}{\underset{\displaystyle H}{C}}-H$$

   b.
$$H-\overset{\displaystyle H}{\underset{\displaystyle H}{C}}-\overset{\displaystyle H}{\underset{\displaystyle \underset{\displaystyle H}{\overset{|}{C}-H}}{C}}-\overset{\displaystyle H}{\underset{\displaystyle H}{C}}-H$$

c.

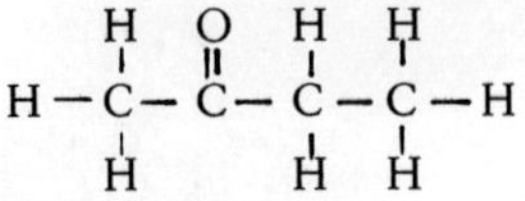

```
     H   O   H   H
     |   ||  |   |
H —  C — C — C — C — H
     |       |   |
     H       H   H
```

d.  ⬡

5. a. Propanone
   b. Ethyne
   c. Butanal

   d. 1–propanol
   e. Butanoic acid

7. a. 1,3–dinitrobenzene
   b. 1–chloro–3–nitrocyclohexane
   c. 1–cyano–3–iodocyclopentane

   d. 1,4–dinitronapthalene
   e. 3–chlorophenol

9. a.

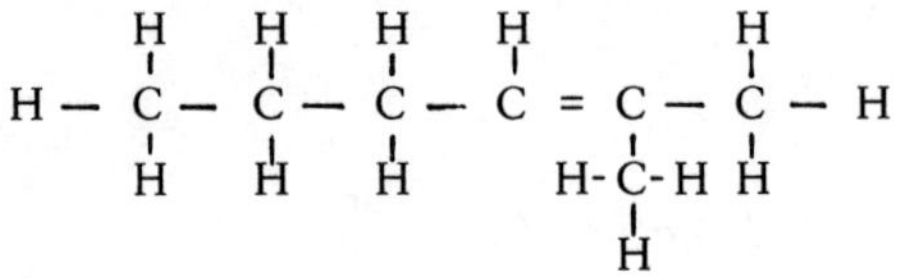

```
       H     H     H     H           H
       |     |     |     |           |
H —    C —   C —   C —   C  =  C  —   C  —  H
       |     |     |           |      |
       H     H     H         H-C-H    H
                               |
                               H
```

   b.

```
                           H              H
                           |              |
       H     H   H  H-C-H      H-C-H   H
       |     |   |    |          |     |
H —    C —   C — C —  C  —   C  =  C  —   C  —  H
       |     |   |    |  H-C-H        |
       H     H   H    H    |          H
                           H
```

   c.

```
     H   H   H   H   H   ⬡   H   H
     |   |   |   |   |       |   |
H —  C — C — C — C — C — C — C — C — H
     |   |   |   |  NO₂  |   |   |
     H   H   H   H      ⬡   H   H
```

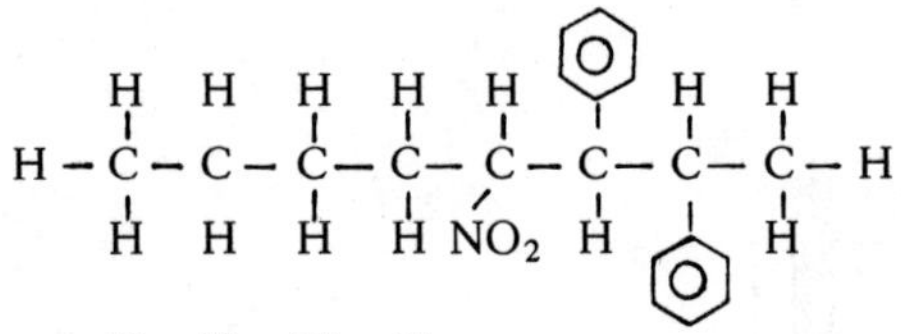

   d.

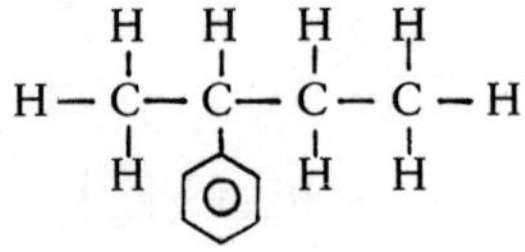

```
     H   H   H   H
     |   |   |   |
H —  C — C — C — C — H
     |   |   |   |
     H   ⬡   H   H
```

   e.

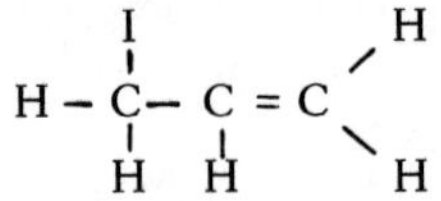

```
     I            H
     |          /
H —  C — C  =  C
     |   |      \
     H   H       H
```

   f.

```
       Cl
       |
Cl —   C — Cl
       |
       Cl
```

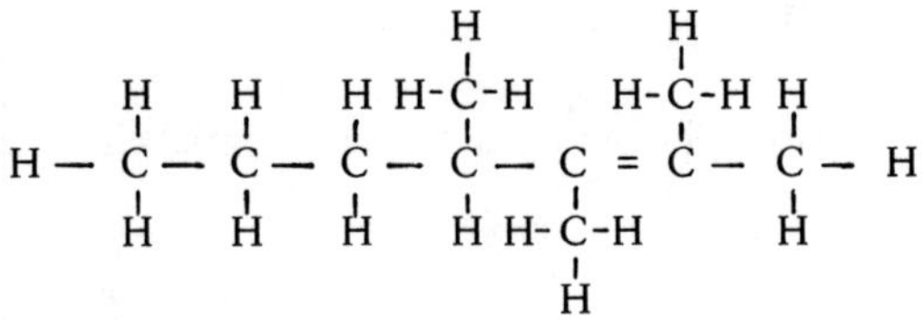

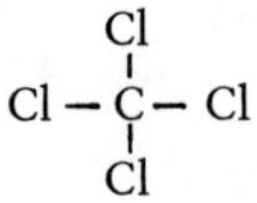

Chapter 16

1.
```
     H  H  H  H  H
     |  |  |  |  |
H —  C— C— C — C— C— H
     |  |  |  |  |
     H  H  H  H  H
```

```
     H     H     H     H
     |     |     |     |
H—   C—    C—    C —   C — H
     |     |     |     |
     H     H  H-C-H    H
                 |
                 H
```

```
           H
           |
     H  H-C-H  H
     |    |    |
H —  C —  C —  C— H
     |    |    |
     H  H-C-H  H
           |
           H
```

3.
```
     H  H  H  H                      H  H  H  H
     |  |  |  |            HNO3       |  |  |  |
H—   C— C— C— C— H          →    H—   C— C— C— C— NO2
     |  |  |  |                      |  |  |  |
     H  H  H  H                      H  H  H  H
```

```
     H  H  H  H
     |  |  |  |
H—   C— C— C— C— H
     |  |  |  |
     H  H NO2 H
```

```
     H  H  H
     |  |  |
H—   C— C— C— NO2
     |  |  |
     H  H  H
```

```
     H  H  H
     |  |  |
H—   C— C— C— H
     |  |  |
     H NO2 H
```

```
     H  H
     |  |
H—   C— C— NO2
     |  |
     H  H
```

```
     H
     |
H—   C— NO2
     |
     H
```

5. a. No reaction   c. Sodium methoxide
    b. Sodium phenoxide   d. Sodium phenoxide

7. a. $CH_3 OH + K_2 Cr_2 O_7 \rightarrow H - C = O$ methanal

    $$\underset{H}{|}$$

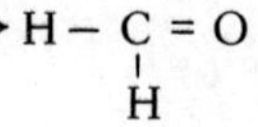

    b. $CH_3 CHOCH_3 + KMnO_4 \rightarrow$ H $-$ C $-$ C $-$ C $-$ H propanone

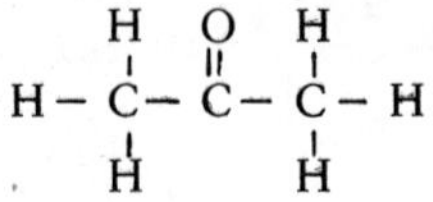

    (with H above and below first C, O double bond above second C, H above and below third C)

    c. $(CH_3)_3 COH + KMnO_4 \rightarrow$ no reaction

9. Butane $\overset{Cl_2}{\rightarrow}$ chlorobutane $\overset{NaOH}{\rightarrow}$ 1$-$butanol $\overset{KMnO_4}{\rightarrow}$ butanal $\overset{K_2Cr_2O_7}{\underset{H_2SO_4}{\rightarrow}}$ butanoic acid

    C-C-C-C $\rightarrow$ C-C-C-C-Cl $\rightarrow$ C-C-C-C-OH $\rightarrow$ C-C-C-C$\overset{O}{\diagdown}$ $\rightarrow$ C-C-C-C$\overset{O}{\diagup}\diagdown$OH

Chapter 17

1. a. Monosaccharide

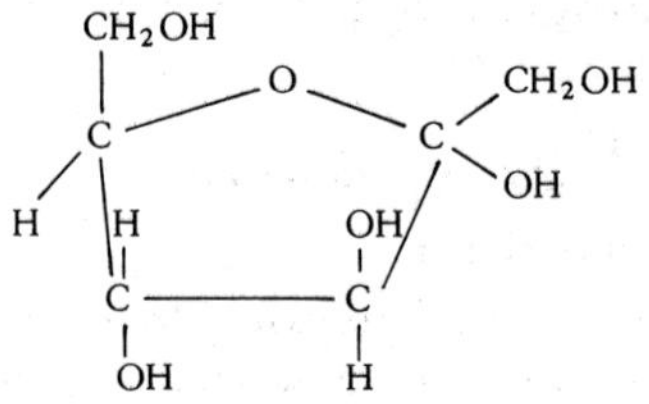

    b. Disaccharide

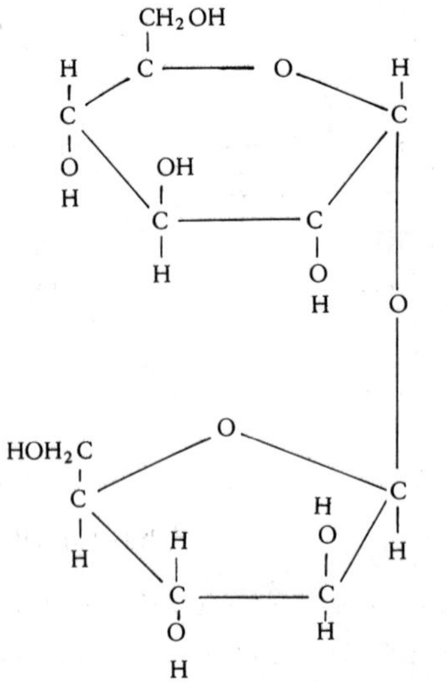

c. Monosaccharide

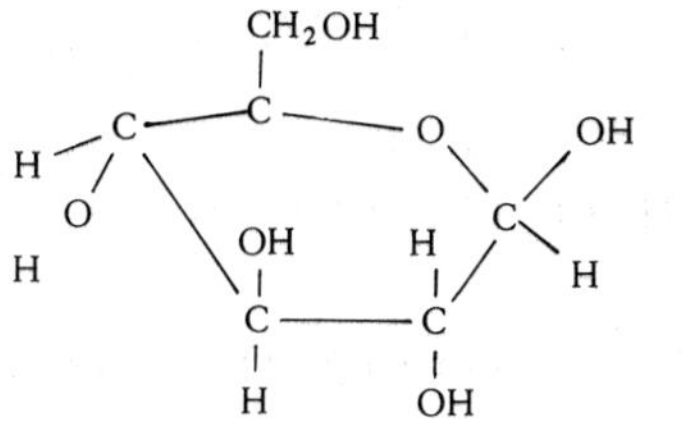

d. Polysaccharide

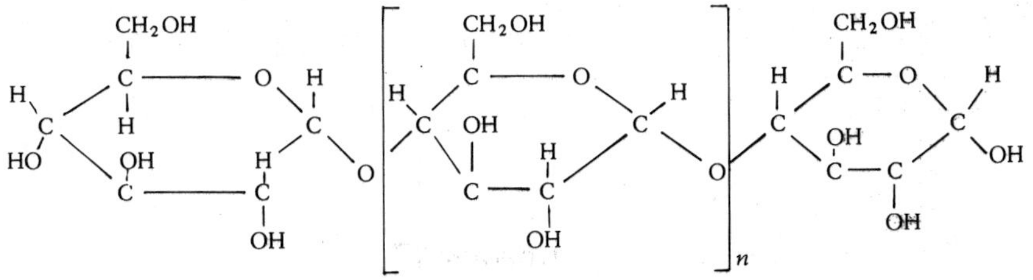

3. It is a polysaccharide and cannot be absorbed by the body until reduced to a monosaccharide.

5. Sucrose $\xrightarrow{H^+}$ glucose + fructose

7. Refer to question 1

9. Both are complex polysaccharides but the difference is the method in which the various monosaccharide groups link together. The linkage is the difference.

11.

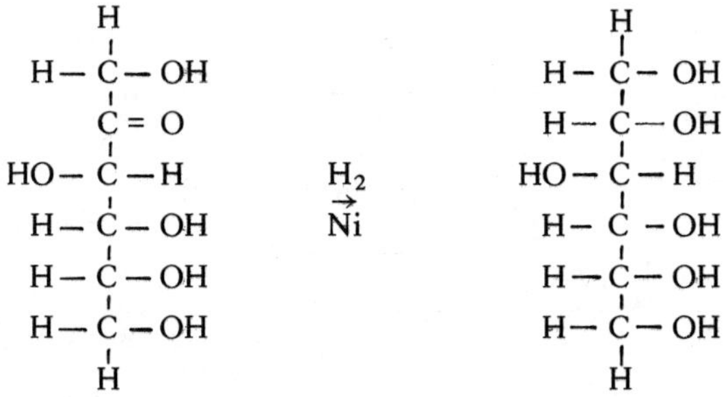

13. See figure 17.8

Chapter 18

1. A fat is a common name for a chemical family called esters.

3. An amino acid is found within a protein. When a protein is hydrolyzed, amino acids are formed.

5. a. $C_2 H_5 COOCH_3 \xrightarrow{H_2} C_2 H_5 CH_2 OH + CH_3 OH$
                                  1–propanol   methanol

    b. $C_3 H_7 COOC_2 H_5 \xrightarrow[C_2 H_5 OH]{Na} C_3 H_7 CH_2 OH + C_2 H_5 OH$
                                      1–butanol   ethanol

    c.

$$
\begin{array}{l}
CH_2 - O - \overset{\displaystyle O}{\underset{\displaystyle \|}{C}} - CH_3 \\[2em]
CH - O - \overset{\displaystyle O}{\underset{\displaystyle \|}{C}} - C_2H_5 + NaOH \\[2em]
CH_2 - O - \overset{\displaystyle O}{\underset{\displaystyle \|}{C}} - CH_3
\end{array}
\rightarrow
\begin{array}{l}
CH_2\,OH \\
| \\
CHOH \\
| \\
CH_2OH
\end{array}
+
\begin{bmatrix}
CH_3\,COO^-Na^+ \\
C_2\,H_5\,COO^-\,Na^+ \\
CH_3\,COO^-\,Na^+
\end{bmatrix}
$$

7.

$$
\sim N - \underset{\underset{R}{|}}{\overset{\overset{H}{|}}{C}} - \underset{\underset{O}{\|}}{\overset{\overset{H}{|}}{C}} - N - \underset{\underset{R_1}{|}}{\overset{\overset{H}{|}}{C}} - \underset{\underset{O}{\|}}{\overset{\overset{H}{|}}{C}} - N - \underset{\underset{R_2}{|}}{\overset{\overset{H}{|}}{C}} - \underset{\underset{O}{\|}}{\overset{\overset{H}{|}}{C}} \sim
$$

(Basic peptide chain, with the R's referring to side-chains.)

Chapter 19

1. Natural radiation occurs in nature whereas artificial radiation is induced by man.

3. Fusion is the combination of nuclei with small masses such as hydrogen. Fission is the process of reducing a heavy nucleus to lighter nuclei. An example is U–235.

5. 174 days

7. $^{208}_{82}Pb$

9. a. To attain the required high temperature
    b. The fusion reaction

Index

tables of, 5-15
volume, 15
Unsaturated hydrocarbons, 192-198, 206, 213-219
Uranium, 2, 246, 247, 250, 251
Uranium disintegration series, 247

Valence:
    bond theory, 57-65
    definition of, 68
    electrons, 57-65, 68
    and outer shell electrons, 57-65
    and oxidation numbers, 65
Vapor, 102-111, 116
Vaporization, 116
Vapor pressure, 106, 110, 116, 263
Voltaic cells, 179, 180
Volume, measurement of, 15, 147-150

Water:
    atomic composition of, 60

boiling point, 116, 117
composition of, 60
density of, 15, 102
electrolysis of, 179, 180
equilibrium with ice, 115, 117 121
freezing point, 115
ionization of, 154, 261
pH of, 154
as a solvent, 115, 128
structure of molecule, 60
vapor pressure of, 263
Water of crystallization, 123-125
Weak acid, 166-168, 261
Weak base, 166-168, 261
Weight:
    definition of, 3
Weight relations in reactions, 95, 142-149

X-rays, 245
    and crystal structure, 123-125
    discovery of, 245